Macroeconomics of North Sea Oil
in the United Kingdom

To Professor Sir Richard Stone

Macroeconomics of North Sea Oil in the United Kingdom

Homa Motamen
Imperial College, London University

in collaboration with Roger Strange
Imperial College, London University

Heinemann Educational Books

Heinemann Educational Books Ltd
22 Bedford Square, London WC1B 3HH

LONDON EDINBURGH MELBOURNE AUCKLAND
HONG KONG SINGAPORE KUALA LUMPUR NEW DELHI
IBADAN NAIROBI JOHANNESBURG
EXETER(NH) KINGSTON PORT OF SPAIN

First published 1983

British Library Cataloguing in Publication Data

Motamen, Homa
Macroeconomics of North Sea oil in the United Kingdom.
1. Petroleum industry and trade—Great Britain—Mathematical models
I. Title
338.2′7282′0941 HD9571.5

Typeset by Text Processing Ltd., Clonmel, County
Tipperary, Ireland and printed by
Biddles Ltd., Guildford Surrey

Contents

Preface

While the question of allocation of oil revenue is relatively new to Britain, it is a familiar and long-standing problem facing the more traditional oil-producing nations. Needless to say, the characteristics of the British economy are markedly different from those of any of the typical oil-based countries. Nevertheless, there are some valuable lessons for Britain to learn from the experience of these countries. At best it can avoid some of the mistakes they have made, for the oil resource, in the British case, as elsewhere, provides an income which is not directly generated from the internal activities of the economic system. Injections of such a revenue into the economy can create imbalances that would not otherwise occur.

A planning model is presented here and used to examine the intertemporal investment strategies of an oil-rich industrially advanced economy such as the United Kingdom. The model is formulated in the framework of Optimal Control theory. The characteristics of the economy after the emergence of the oil sector are expressed first by means of a macroeconomic model which is then applied to a dynamic optimization process. A set of analytical solutions are derived which are subsequently verified numerically.

This volume is based on the findings of a research project financed by the Social Science Research Council. The research assistants who contributed at different times to this project were Roger Strange, Stephen Hall, Brian Price and Sonia Brunsdon to whom I am very grateful. I would like to thank the staff of H.M. Treasury and Inland Revenue for their valuable time in discussing details of our calculation of the government oil revenue. I am also grateful to Dr D.A.Livesey who acted as an external member of the project. I have benefited from stimulating discussions with Michael Posner, Len Brookes, Professor Colin Robinson, Professor Aubrey

Silberston, Dr Christopher Bliss, Alex Kemp, and Dr Paul Eisenklam. I am grateful to Dr Terry Barker and John Kay for their comments. My special thanks to Patricia Burge whose role as the project secretary was invaluable. Her patience and conscientious work made the preparation of this manuscript possible. I am also grateful to Mrs Joan Wright for providing extra typing assistance. I would like to express my gratitude to my husband, Brian Scobie, who gave me a great deal of help and support to complete the research project.

Homa Motamen
Director,
North Sea Oil Project

May 1982

List of Tables and Figures

1 Introduction

This book is concerned with the macroeconomics of North Sea oil depletion coupled with recycling of the oil revenue in the UK. The discovery and the subsequent development of this resource is a phenomenon which will have lasting repercussions on the British economy. The question of how best to spend the proceeds of North Sea oil is clearly a long-run problem. Ironically most of the policies proposed and debated in the national press[1] as well as those put forward in the Government White paper, 'The Challenge of North Sea Oil',[2] are primarily short-run. It could be that the nature of the political system in Britain prevents the implementation of long-term plans. With governments changing every three or four years, policies on the whole tend to be short-run.

In order to formulate an effective intertemporal plan to utilize North Sea oil revenue, it is vital to look at the lifetime of the resource as a whole, and ask the question: what does Britain wish to achieve by the time the resource is exhausted? It is inevitable that North Sea oil will leave a mark on the face of the British economy, however small its share of GNP. Thus, depending on how the receipts from the oil sector are managed, the resource can be beneficial or, indeed, quite damaging and capable of paralysing some of the self-equilibrating forces in the economy. To avoid the latter, it is crucial both to treat the resource as capital and not income, and to plan carefully the priorities for long-run investment before it is frittered away on non-productive projects.

A useful method of studying the problem is the method of Optimal Control, which is a technique of optimum choice over time. The main advantage of this technique over others lies in the fact that it provides a unified approach to the problem. This method is applied here to study depletion strategies in conjunction with expenditure policies of North Sea oil revenue, and to analyse their impact on the different macroeconomic variables.

The procedure for carrying out this exercise involves two stages. First, a macroeconomic model of the UK economy is built. Next, it is adapted so that it can be applied to the optimal control technique. This is done in chapter 2. The macro model is a fairly aggregate one, comprising seventeen equations, which explain the overall structure of the British economy. It is an open-economy model designed to bring out the role of the newly-developed oil sector and to highlight its potential in influencing directly (through the revenue side) and indirectly (through the balance of payments side) the rest of the economy.

The objective function is expressed in the form of the following question: what is the maximum global wealth that can be achieved, through the intertemporal allocation of the revenue, by the end of the life of the oil? This objective function is chosen mainly because it results in maximizing the long term consumption of the economy. As Forsyth and Kay[3] have pointed out 'it has become all too common to believe that the end of economic activity is production rather than consumption. This seems to be born of spurious moralizing and the power of interest groups of producers'. So it is important to realize that the fundamental object of the economy is not maximization of national wealth for its own sake. Rather it is maximization of the consumer's standard of living. Higher consumption is caused by a higher permanent income stream, which in turn results from an increased level of total assets.

A set of primary solutions to this planning model are first received. These are reached through a simultaneous solution of the problem. Since the model is an open economy one and the objective is global maximization of wealth, this implies that the economy has a choice between investing at home and overseas. Moreover, since the problem is a dynamic one, the proper allocation of the resource involves determining when and where to invest, that is, when to invest overseas, and when at home. Hence, the problem of absorption capacity of the economy for a newly-acquired income, namely that of oil, becomes relevant. This can be achieved through the method of optimal control where the allocation of the resource between the home and overseas sectors is checked against the final target.

The analytical solutions which emerge are such that when the balance of payments constraint first becomes binding[4] it is optimal to invest more overseas even if it involves asking for a lower rate of return abroad. Conversely, when the constraint is nearly removed, it is optimal to invest more at home by asking for a lower rate of return. At the same time the solutions indicate over the period when the balance of payments constraint is binding, there should be a cut-back in oil production. When the constraint is fully removed, the optimal rule for oil depletion turns out to be a modified version of the Hotelling's rule. The solutions show that where the state is the oil producer and the objective is maximization of non-oil wealth over the life of oil, the extraction rate should be such that, in equilibrium, the rate of change of oil prices should be less than the going rate of interest. This is lowered by a coefficient which is itself a combination of the average propensity to consume, the rate of income-related taxes and the rate of expenditure taxes in the economy. The rationale for these solutions are discussed in the text.

In chapter 3, some of the major macroeconomic models of Britain are examined and a brief summary of each model is included. This has two objectives: (i) to show that the basic nature of these models tends to be short term, and moreover, that none of them determine any depletion policies for the North Sea oil[5]—hence, the necessity to formulate a new model; (ii) to provide the information needed to quantify the parameters of our model, later, in chapter 5. Six main large econometric models of the UK are considered. These are the models of Her Majesty's Treasury, the National Institute, the Cambridge Growth Project, the London Business School, the Bank of England and the Cambridge Policy Group.

Chapter 4 contains a detailed description of the actual oil taxation system and its historical evolution, and includes some empirical estimation of the likely magnitude of government revenues accrued from the North Sea oil taxation. In chapter 5, first, the oil taxation system during 1982 is described, as well as the subsequent arrangements which take effect from 1 January 1983. Second, the changes that the system has undergone since its inception are outlined, and third, the magnitude of the revenue is measured for the years 1976–2000 (which coincides with our planning period in this book) for the twenty-four fields whose proven recoverable reserves serve as a basis for the initial oil stock[6] assumed in our optimization calculations.

The material of chapter 4 has been prepared in such a way that it is as up-to-date as possible. This has been necessary because the oil taxation in Britain seems to be subject to continual revisions! Moreover, some of the significant variables such as the price of oil, rate of dollar/sterling exchange rate, rate of oil extraction and the quantity of proven reserves can change as a result of different market and macroeconomic conditions. The latest taxation change which occurred before this book went to press was announced in the March 1982 Government Budget by the Chancellor of the Exchequer. Hence, chapter 4 was updated to incorporate these specific changes. When revising this section, however, we also used the latest figures for all the other variables which entered into our computations of the government revenue. These included the estimate of the proven oil reserves (which is slightly less than the original figure used in chapter 5 as our stock of oil reserves), the exchange rate and the price of oil.

The price of oil can be extremely unstable and whatever projections one uses may prove to be at variance with reality. The erratic behaviour of the oil market can radically alter the industry's expectations of the future, whereby its predictions are highly influenced by the events of the present. However, it is important not to be too influenced by short-term fluctuation when projecting oil prices over a long period, i.e. as long as twenty years. Moreover, it should be emphasized that the projections invoked in this study are utilized primarily in order to demonstrate a methodology in macroeconomic planning for an oil producing country. Sensitivity analyses are conducted at all levels in order to reveal the effect of variations in the price path of oil.

The next stage of the book focuses on the application of the model, illustrating how the analytical solutions reached formally, in chapter 2, can be verified numerically. However, in order to apply the model and obtain numerical values for the optimal trajectories of investment and oil depletion, one needs to supply it with a certain amount of initial data. Such data comprises the values of the parameters of the model and the constraints, the projected time path of the exogenous variables and the values of the stock variables in the starting year.

Chapter 5 is devoted to the preparation of the data for the utilization of the model. The values of the model parameters are quantified by means of economic reasoning. They are not determined by applying conventional econometric estimation techniques. The model parameters and the initial stock values are estimated by considering the theoretical restrictions and existing econometric works. This is done for two reasons. Firstly, the experience of the past may not be necessarily relevant to the future, particularly with the emergence of the oil sector—the latter can, of itself, generate economic changes which were not present in the pre-oil period. Secondly, it is felt that the combination of some of the existing econometric work and understanding of the theoretical underpinnings of the macroeconomic system would yield a more robust model than a new set of direct econometric estimates. Wherever possible, of course, reference is made to the existing literature and to some of the econometric results in other studies. Various sensitivity analyses are carried out by varying the size of each parameter and studying the effect on the overall solutions. Another reason for adopting this method of estimation is related to the nature of our study, which is primarily a conceptual study. The research is designed to demonstrate a methodology in planning, applying optimal control theory. Thus it is important to emphasize that it is not so much the absolute values of the solutions that matter. Rather, the relative size of the figures associated with the numerical solutions should be used to understand the implications of the results.

Since the model formulated here is essentially a real model, all the numerical calculations are done in constant prices, and all the variables expressed in constant 1975 prices. The time period considered for this planning exercise is two-and-a-half decades, commencing in 1975—the year that marks the advent of oil revenue in Britain.

Chapter 6 deals with the numerical results obtained from the computer programme for our planning problem. The optimal paths for the control variables which are the non-oil domestic investment and the rate of oil extraction are determined. These paths are then analysed in order to understand the underlying interplay of the relationships involved. The computational method is based on a non-linear programming technique, called the methods of feasible directions. The optimal solutions are reached through a convergence sequence. The advantage of this approach is that it can be applied to non-linear models with a number of inequality constraints.

In chapter 7 the numerical results for the basic model[7] are outlined. The main purpose of this chapter is to see how the optimal solutions might change if some of the basic parameters were varied. This exercise is particularly useful when considering such variables as government expenditure, the income tax rate or the expenditure tax rate. All these variables are to some extent under government control and changes in them could affect oil depletion policies and investment of the revenue. It is important, therefore, to know precisely what effects alterations of these variables would have. Further simulations against the basic model are conducted and compared with the solutions outlined in the previous chapter. In each simulation only one parameter or relationship is varied at a time. This allows us to investigate the effect of each change, and to examine how robust the basic model solutions are to variations in the basic model structure.

Finally, chapter 8 summarizes the main results of the study together with their relevant policy implications. It should be pointed out that this book is written in such a way that the reader not equipped with mathematical expertise can easily understand the underlying theme and concepts. Chapters 2, 5 and part of chapter 6 (sections 6.1—6.4) may be omitted without much loss of understanding of the basic principles. The non-technical reader can move straight to the section 2.5 of chapter 2 and proceed to chapters 3 and 4, followed by chapter 6, section 6.5 (where the economic implications of the numerical solutions are discussed), and continue to the end of the book.

2 The Model and Its Analytical Solutions

2.1 Introduction

There seems to be a tendency to determine the rate of oil extraction independently of the decisions to invest the proceeds of the resource. In this chapter a general framework is set out, whereby these two important policy variables can be analysed jointly. The methodology proposed is one of optimal control theory combined with a long-run macroeconomic model. The model developed here is structured with special reference to the UK economy but it can be applied, with the necessary modifications, to any industrialized economy having a flexible exchange rate.

The analytical solutions to this planning model are shown in this chapter. They are determined through a simultaneous solution of the problem. The model is an open economy one and the objective is global maximization of non-oil wealth, which in turn implies that the economy has a choice between investing at home and overseas. Since the problem is a dynamic one, the proper allocation of the resource involves determining when and where to invest, that is, when to invest overseas and when to invest at home. The problem of absorption capacity of the economy for a newly-acquired revenue, namely oil, becomes relevant. This can be achieved through the method of optimal control where the allocation of the resource between the home and overseas sectors is checked against the final target.

2.2 Structure Of The Macroeconomic Model Of The UK

The macroeconomic model formulated here is a fairly aggregate one, comprising seventeen equations, which explain the overall structure of the British economy. It is an open-economy model designed to bring out the role of the newly-developed petroleum sector and to highlight its potential in influencing directly and indirectly the rest of the economy. The model is essentially a long-run model, given the specific nature of the problem posed in this study. The time period under consideration is roughly two-and-a-half decades, which it is assumed to span the life of North Sea oil. There are altogether seven identities and ten behavioural equations in the macroeconomic model and we shall explain each equation individually (see Tables 2.1 and 2.2).

The economy is broken into three main sectors: oil, non-oil domestic and non-oil overseas. Although the size of the oil sector is substantially smaller than that of the non-oil, it was necessary to draw the distinction in order to carry out a planning exercise for the proceeds of the resource. Equation 2.1 which is the first of the identities, gives the national income as the sum of output from these sectors. More specifically, gross national income at market prices, N_t, comprises:

(i) gross domestic non-oil value-added at factor cost, D_t, plus expenditure taxes on non-oil output, l_t^{11}, to give gross domestic product at market prices;

(ii) value-added in the oil sector at market prices, P_t; and

(iii) net income from overseas, F_t.

The reasons for expressing P_t at market prices will become more clear after equation 2.4 has been explained.

Equation (2.2) depicts an aggregate production function of the economy outside the petroleum sector. The non-oil domestic output, D_t, is shown as a non-linear function of the lagged values of capital, K_t, and labour, L_t. It is assumed that in the long run the production function exhibits decreasing returns to scale with respect to the capital stock variable. This is portrayed by restricting the value of the exponents of the capital stock to less than one. This restriction on the exponent of the capital variable is imposed here, for it is believed to characterize the UK economy best, in the light of the many obstacles it faces in the way of production, such as work stoppages at different production levels and various industrial and labour problems. As a result, *ceteris paribus*, the proportional increases in output decline as more of the capital input is added. Technical progress is also built into the production function and is formulated as an increasing function of time.

Raw materials used in the domestic production are postulated to be proportional to the level of current non-oil domestic output. This is shown in equation 2.3 whereby as output rises the demand for raw material increases. The value of output generated in the oil sector, P_t, is given by equation 2.4 as the product of the oil price, π_t, and the quantity of oil produced, q_t. The price of oil is taken as an exogenous variable.

Since the UK is not a member of the price-setting cartel, OPEC, it is thought best to regard the economy as a 'price-taker'.[1] Equation 2.4 also has an added identity attached to it which splits total oil revenue into two parts:[2] the public sector share of oil, l_t^1, and the private sector, namely the companies' share of oil P_t^c.

While q_t is a flow variable determining the rate of extraction of crude oil, Q_t is the stock variable which measures the volume of oil reserves beneath the North Sea floor. Equation 2.5 is a first order difference equation which gives the volume of production in each time period as the change in reserves over two consecutive time periods. So the volume of reserves in the last time period, Q_{t-1}, less the level of oil extracted this period, together determine the current volume of reserves remaining in the sea bed, Q_t. With a fixed level of reserves, Q_0, this would imply that if there is any production in a given time period, say t, by definition Q_t would be larger than Q_{t+1}. We shall explain later how q_t is used as a control variable and Q_t as a state variable in our Optimal Control model.

The public sector share of oil revenue comprises royalties, corporation tax, supplementary petroleum duty[3] and petroleum revenue tax (PRT). This makes the government revenue from this sector an increasing function of oil output.[4] The relationship is presented in equation 2.6.

Gross domestic fixed capital formation, I_t, is expressed as changes in capital stock from one period to the next—equation 2.7. This is shown by means of a first order difference equation. Capital stock, K_t, and investment are both measured as 'gross' to avoid complications over calculation of the appropriate depreciation rates. Not only has estimation of depreciation been a controversial and unresolved question, it is also of no direct relevance to the problem we are studying. It should be noted that the variable I_t embraces investment in the non-oil sector only and excludes any capital expenditure in the petroleum sector. Moreover, I_t applies essentially to productive investment. This implies that the flow of investment in our model comprises two specific categories:

(a) *direct investment* to establish productive capacity and raise total domestic output, such as investment in manufacturing plants or agricultural machinery;

(b) *investment on productive infrastructure* to facilitate the expansion of direct investment. This serves indirectly to raise productive capacity, for example through construction of more roads, electricity plants, renovating port facilities and warehouses.

Investment on social infrastructure is not, however, included in I_t. Expenditure to meet social needs, such as hospitals and schools, are included in the item public sector expenditure, G_t. This classification is used here on the argument that although social infrastructure may be a prerequisite of productive investment, its existence does not necessarily imply that productive investment will be induced. Therefore, the incremental output-capital ratio in this model is applicable only to items (a) and (b) above and not to social infrastructure.[5]

Investment in the oil sector is treated separately. It is assumed as exogenous to the model, for the firms investing in the North Sea are mainly multinational companies. Their capital, accordingly, is not necessarily UK owned, and is supplied from their profits made in different parts of the world. Such capital would be regarded as part of their international assets. Thus capital expenditure in the petroleum sector enters the model as an exogenous variable through equation 2.6 where Petroleum Revenue Tax is computed. Here the companies' investment expenditure is offset against their revenue in calculating their tax liabilities to the government.

Equation 2.8 gives net income from abroad, F_t, as a function of the portfolio of assets overseas during the previous period, E_{t-1}. Overseas income should rise in constant proportion, θ, to overseas assets in the previous period. E_t in this context includes debts, which would be negative assets. Since the concept of 'net' is applied to overseas assets, E_t, then F_t, accordingly, can take positive or negative values.

The net change in foreign assets from one period to the next is shown in equation 2.9 as the balance of payments on the current account, B_t. Although this assumption may not necessarily hold in the short run, it is perfectly valid in the context of our model, as we are considering the relationship over a long period.

Equation 2.11 depicts a simple long-run consumption function which gives aggregate private consumption as dependent on disposable income. The latter is defined as the sum of domestic non-oil income, less income-related taxes, l_t^{111}, and net income from abroad. This formulation is based on the classic permanent income hypothesis which conforms with the long-run nature of the model. It could be argued that the consumption function should include the revenue from the oil sector, P_t, as one of its explanatory variables. However, although the oil sector may affect the size of the aggregate private consumption indirectly, because of its peculiar characteristics it has little direct influence. The components of P_t are shown in equation 2.4 to consist of the public sector share, l_t^1, and the oil companies' share, P_t^c. Due to the multinational nature of the oil companies, most of their revenue tends to flow abroad and to become distributed among their various bases overseas. The only accountable part of this revenue that would remain in the UK would be payments to British labour employed and payments to the non-oil UK industries providing inputs to the North Sea oil industry. The former category of payments is so small in relation to the total petroleum companies' revenues that its contribution is likely to be negligible as the North Sea oil industry is primarily capital intensive. The latter category of payments, though small, is likely to have more of an indirect effect and its inclusion in the consumption function is likely to cause greater inaccuracy than its omission.

The volume of income-related taxes l_t^{111}, and that of the expenditure tax, l_t^{11}, are determined through equations 2.12 and 2.13 respectively. The former is directly proportional to the sum of the values of non-oil domestic income and non-oil overseas income.[6] This fixed proportion is shown through the coefficient ε. The extent of expenditure tax is directly related to the size of aggregate private consumption. This is shown by means of coefficient σ.

Equation 2.14 is simply a rearrangement of the national income accounting identity. It defines the balance of payments on the current account, B_t, as the residual of net national income minus aggregate expenditure, namely, consumption, investment and government expenditure, G_t. Suffice it to say that the components of expenditure constitute aggregate domestic demand. It should also be noted that B_t is a composite term of visible and invisible exports and imports.

Equation 2.15 is another identity, which defines total imports, M_t, as the difference between total exports, X_t, and the balance of payments on the current account. Total exports are then further subdivided into non-oil exports, V_t, and the share of oil which is exported. The value of oil exported is expressed as a proportion, ω, of total UK oil produced, P_t. It should be noted that the value of oil exported is a 'net' concept, and it is net of the foreign oil imported in the UK. ω is taken to be a constant parameter over the time horizon under consideration. This may not be a true situation and the value of net exports may change during the life of the North Sea oil. Therefore, although in the first instance we will consider a constant parameter, fluctuations in this parameter can be examined in the context of this model.

Equation 2.16 is a behavioural equation expressing non-oil exports as a non-linear function of the last period's exchange rate, S_{t-1}. The underlying assumption behind this relationship is that non-oil exports move inversely with the exchange rate, so that as the latter rises in any period, the quantity of the former falls in the next period and vice versa. It is thus postulated that the first derivative of this equation is negative.

Movements in the exchange rate have in turn been related to the volume of oil revenue. Equation 2.17 determines this relationship through a non-linear function with a positive first derivative and a negative second derivative. It is assumed to have a fairly substantial positive intercept on the basis that clearly the exchange rate will be positive even when total oil revenue is zero.

In constructing this dynamic model it is assumed that investment is correctly allocated to match the economy's pattern of demand. It was pointed out before that, to study the planning strategy for North Sea oil revenue, we used the method of Optimal Control. The procedure for applying this method involved two stages. First a macroeconomic model of the UK economy had to be built, which we have already explained. Second the problem had to be formulated in such a way that it could be applied to the technique of Optimal Control. In other words an explicit economic target and certain intertemporal constraints had to be expressed.

Table 2.1 Notations used in the macroeconomic model

(i) *Variables*

N_t = Gross national income
D_t = Value added in the non-oil sector at factor cost
P_t = Value added in the oil sector, market prices
F_t = Income from abroad
K_t = Fixed capital stock gross
I_t = Gross non-oil fixed capital formation
L_t = Labour input
R_t = Raw materials used in domestic production
π_t = International price of crude oil

Q_t = Stock of oil reserves
q_t = Quantity of crude oil extracted annually from North Sea
l_t^1 = Total government take from North Sea oil
l_t^{11} = Expenditure tax
l_t^{111} = Income-related tax
P_t^c = Company share of value-added of oil
E_t = Stock of assets abroad
C_t = Personal consumption
B_t = Current account balance (changes in reserves)
G_t = Government expenditure
X_t = Total exports
V_t = Non-oil exports
S_t = Effective exchange rate
h_t = Rate of change of effective exchange rate

(ii) *Parameters*
g = Scale variable in the production function
β = Rate of technical progress
α = Elasticity of response for capital
γ = Elasticity of response for labour
ξ = Raw materials used as a proportion of current output
θ = Yield on investment overseas
η = Average propensity to consume
ε = Coefficient of income-related tax
ω = Proportion of oil exported
σ = Coefficient of expenditure tax
m_0 =, m_1 = } Parameters of the export equation
Ω_0 =, Ω_1 =, Ω_2 = } Parameters of the foreign exchange equation
ϕ = Essential imported share of raw materials used in non-oil production
μ = Essential imported share of aggregate comsumption
v_0 = a combination of parameters σ, η and ε as shown in equation 2.19 defined to facilitate exposition of argument.
v_1 = a combination of parameters σ, η, ε and μ as shown in equation 2.21
$v_2 = v_1 + \phi\xi$ as shown in equation 2.22

Table 2.2 The macroeconomic model

$N_t = D_t + l_t^{11} + P_t + F_t$	(2.1)
$D_t = g(1+\beta)^t K_{t-1}^{\alpha} L_{t-1}^{\gamma}$	(2.2)
$\beta \geqslant 0 \quad 0<\alpha<1 \quad 0<\gamma<1$	
$R_t = \xi D_t$	(2.3)
$P_t = \pi_t q_t \equiv l_t^1 + P_t^c$	(2.4)
$q_t = Q_{t-1} - Q_t$	(2.5)
$l_t^1 = f(q_t)^*$	(2.6)
$I_t = K_t - K_{t-1}$	(2.7)
$F_t = \theta E_{t-1}$	(2.8)
$\theta > 0$	
$E_t = B_t + [E_{t-1}/(1+h_t)]$	(2.9)
$h_t = \dfrac{S_t - S_{t-1}}{S_{t-1}}$	(2.10)
$C_t = \eta(D_t - l_t^{111} + F_t) \equiv \eta(1-\epsilon)(D_t + F_t)$	(2.11)
$l_t^{111} = \epsilon(D_t + F_t)$	(2.12)
$l_t^{11} = \sigma C_t$	(2.13)
$B_t = N_t - C_t - I_t - G_t$	(2.14)
$M_t = X_t - B_t \equiv V_t + \omega P_t - B_t$	(2.15)
$V_t = m_0 S_{t-1}^{m_1}$	(2.16)
$\dfrac{dV_t}{dS_{t-1}} < 0$	
$S_t = \Omega_0 + \Omega_1(\pi_t q_t)^{\Omega_2}$	(2.17)
$\dfrac{dS_t}{dq_t} > 0$	

* The formula for total government take from the oil sector is applied.

The objective function was formulated to answer the following question: What is the maximum global wealth that can be achieved, through the intertemporal allocation of the revenue, by the end of the life of oil? A fixed time horizon is considered initially, which can become variable later depending on new discoveries or changes in depletion rate. By 'global wealth' is implied wealth accumulated both within and outside the United Kingdom. Moreover, the concept of wealth here refers to productive wealth, i.e. income-generating assets.

A dynamic objective is expressed as the sum of domestic and foreign investment over the life of the resource to be maximized. It seems appropriate at this stage to explain why such an objective is chosen.

The functional relationships of the model are specified such that domestic output is a function of domestic capital, and overseas income is a function of the size of overseas assets. Also, private consumption is directly related to the size of the two incomes less taxes, namely, disposable income. Therefore, given the positive relationships among these variables, maximizing the sum of global assets would be the same as maximizing global income or maximizing aggregate consumption. The variables are effectively interchangeable.

It is worth stressing that our model objective does in fact encompass most of the targets expressed in the 1978 UK Government's White Paper 'The Challenge of North Sea Oil'. The four main goals listed there were: investing in industry, improving industrial performance, investing in energy and increasing essential public services. Achievement of these national aims in the long run would be ensured through the objective function expressed in our model. Presumably the wish of the government at the time was encompassed in a further objective such as raising the level of national income, raising exports and so on, and not investment in industry for its own sake. These future ambitions would come under the umbrella of dynamic targets in our planning model.

Another important feature of the objective function specified here is its treatment of the social rate of time preference. Conventional discounting (which is equivalent to different weights for different time periods) is not applied here, because the main distinction in our study lies between the 'oil generation' and the 'post-oil generation'—the reason being that the life of North Sea oil is regarded as short enough to be limited to basically one generation. So by maximizing global assets over the life of the resource, one is in effect maximizing global wealth and, hence, the aggregate consumption of the oil generation. The question then arises as to what happens after period T, when the oil is supposedly exhausted. If the planning period is extended beyond period T, should one adopt discounting and introduce some social rate of time preference? Here the post-oil generation is regarded as basically as important as the oil generation, and the former not discounted in favour of the latter on the grounds that the post-oil generation is not endowed with the 'god-send' of the resource that the oil generation is. In short, within the oil generation, we do not distinguish between income (and/or consumption) at different time periods. We maximize the life-cycle consumption when the life of North Sea oil is assumed to span, basically, over one generation.

Next the intertemporal constraints had to be specified. Given that the model is a fairly aggregate one and designed to tackle a long-run problem, we consider two basic constraints. The first is a balance of payments constraint which postulates a minimum imports requirement. In each time-period the economy is in need of importing a certain amount that is necessary for domestic production and consumption. Such imports are categorized as 'essential'. Total imports may well exceed essential imports, in which case the additional imports are regarded as 'non-essential'. The purpose of introducing this constraint into the model is to illustrate that there is an upper limit to the maximum surplus the country can enjoy at any given time, or alternatively a lower limit below which imports cannot fall. The constraint is expressed as an inequality constraint whereby the balance of payments on the current account is less than or equal to total exports minus 'essential' imports. 'Essential' imports are presented as a proportion, ϕ, of total raw materials used in domestic non-oil production plus a proportion, μ, of private consumption.

The other constraint states, basically, that the size of the oil reservoir is finite and cannot be further expanded. The country can use up all the oil over the time horizon considered, or have some left in the ground by the end of the planning horizon. Effectively, the model has one terminal constraint and an intertemporal constraint in each period.

For the problem to be adapted into an optimal control framework, we have to determine what are called the 'control variables'. Two control variables are defined here: the depletion rate of North Sea oil and the rate of domestic investment. The latter is specified as non-oil gross domestic fixed capital formation.

Finally, to determine the optimal policies for achieving the maximum wealth when the resource is exhausted, the model has to be expressed in terms of state, control and exogenous variables. There are three state variables; (i) the stock of non-oil domestic capital, (ii) the stock of overseas assets; and (iii) the stock of oil reserves in the North Sea. The state variables determine the state of the system in each time period. The exogenous variables of the model are: (i) price of oil, (ii) labour employed in the domestic non-oil production, and (iii) government expenditure.

In order to solve the optimal control problem, a reduced form of the macroeconomic model is obtained and both the reduced form and the constraint, as well as the objective function, are expressed in terms of the state, control and exogenous variables. The next consideration is explicit dynamic formulation and analysis of the optimization and programming problem. Having obtained the solutions, we then explain their properties. The main tool of analysis used here is an examination of the first order conditions for a maximum.

The optimization problem can be formally stated as:

$$\text{Max} \sum_{t=1}^{T} I_t + \sum_{t=1}^{T} B_t$$

Subject to

$$B_t \leqslant X_t - \phi R_t - \mu C_t \qquad (2.18)$$

$$Q_T \geqslant 0$$

$$I_t \geqslant 0 \qquad q_t \geqslant 0$$

The state variables are K_t, E_t and Q_t. The control variables are I_t and q_t; and the exogenous variables are π_t, L_t and G_t. The remaining variables are endogenous.

2.3 Reduced Form Of The Model

The reduced form of the model is obtained by expressing the entire model, that is, the equations of motion and the constraints, in terms of the state, control and exogenous variables.

(i) *Equations of motion*

There are three equations of motion, one for each state variable

(a) The first equation of motion, related to the domestic non-oil capital, is the same as equation 2.7 of the macromodel.

$$K_t = K_{t-1} + I_t$$

(b) The second equation, related to the portfolio of overseas assets, is derived in the following way:

First, using equation 2.4 and substituting for N_t and C_t from equations 2.1 and 2.11, gives

$$I_t = D_t + l_t^{11} + P_t + F_t - B_t - \eta(D_t - l_t^{111} + F_t) - G_t$$

Now, substituting for B_t using equation 2.9 and for l_t^{11} and l_t^{111} using equations 2.13 and 2.12,

$$I_t = D_t + \sigma\eta[D_t - \epsilon(D_t + F_t) + F_t] + P_t + F_t - E_t + \frac{E_{t-1}}{1+h_t} - \eta D_t + \epsilon\eta(D_t + F_t) - \eta F_t - G_t$$

Rearranging this equation such that

$$I_t = (1 + \sigma\eta - \epsilon\sigma\eta - \eta + \epsilon\eta)D_t + (1 + \sigma\eta - \sigma\epsilon\eta + \eta\epsilon - \eta)F_t + P_t - E_t + \frac{E_{t-1}}{1+h_t} - G_t$$

and expressing it in terms of E_t, the portfolio of overseas assets, and then rearranging once again,

$$E_t = -I_t + (1 + \sigma\eta - \epsilon\sigma\eta - \eta + \epsilon\eta)(D_t + F_t) + P_t + \frac{E_{t-1}}{1+h_t} - G_t$$

For convenience we define

$$(1 + \sigma\eta - \epsilon\sigma\eta - \eta + \epsilon\eta) \equiv \upsilon_0 \qquad (2.19)$$

and then substitute for P_t using equation 2.4. Thus:

$$E_t = -I_t + \upsilon_0(D_t + F_t) + \pi_t q_t + \frac{E_{t+1}}{1+h_t} - G_t$$

Finally, by substituting for D_t and F_t using equations 2.2 and 2.8 we can express E_t in terms of the state, control and exogenous variables, as follows:

$$E_t = -I_t + \upsilon_0[(1+\beta)^t K_{t-1}^{\alpha} L_{t-1}^{\gamma}] + \upsilon_0(\theta E_{t-1}) + \pi_t q_t + \frac{E_{t-1}}{1+h_t} - G_t \qquad (2.20)$$

(c) The third equation is the same as equation 2.5

$$q_t = Q_{t-1} - Q_t$$

(ii) *Constraints*

The constraints also have to be expressed in terms of state, control and exogenous variables.

(a) The first constraint,

$$B_t \leq X_t - \phi R_t - \mu C_t$$

is the balance of payments constraint. This can be expressed in terms of the state, control and exogenous variables in the following way:

Using equations 2.14 and 2.1 to replace B_t and N_t, we obtain

$$D_t + l_t^{11} + P_t + F_t - C_t - I_t - G_t \leqslant X_t - \phi R_t - \mu C_t$$

Substituting for l_t^{11} from equation 2.13 and for C_t and R_t from equations 2.11 and 2.3 respectively, we obtain

$$D_t + \sigma\eta(1-\epsilon)(D_t + F_t) + P_t + F_t - \eta(1-\epsilon)(D_t + F_t) - I_t - G_t \leqslant X_t - \phi\xi D_t - \mu\eta(1-\epsilon)(D_t + F_t)$$

Rearranging,

$$[1 + \sigma\eta(1-\epsilon) - \eta(1-\epsilon) + \phi\xi + \mu\eta(1-\epsilon)]D_t + [1 + \sigma\eta(1-\epsilon) - \eta(1-\epsilon) + \mu\eta(1-\epsilon)]F_t + P_t - I_t - G_t - X_t \leqslant 0$$

Defining

$$[1 + \sigma\eta(1-\epsilon) - \eta(1-\epsilon) + \mu\eta(1-\epsilon)] \equiv \upsilon_1 \qquad (2.21)$$

and

$$\upsilon_1 + \phi\xi \equiv \upsilon_2 \qquad (2.22)$$

and substituting for P_t and X_t using equations 2.4, 2.15, 2.16 and 2.17

$$\upsilon_2 D_t + \upsilon_1 F_t - I_t - m_0[\Omega_0 + \Omega_1(\pi_{t-1}q_{t-1})^{\Omega_2}]^{m_1} + (1-\omega)\pi_t q_t - G_t \leqslant 0$$

Finally substituting for D_t and F_t using equations 2.2 and 2.8 of the macro model, the constraint can be expressed in terms of the state variables (K_t and E_t), control variables (I_t and q_t) and exogenous variables G_t, π_t and L_t.

$$\upsilon_2[g(1+\beta)^t K_{t-1}^{\alpha} L_{t-1}^{\gamma}] + \upsilon_1(\theta E_{t-1}) - I_t - m_0[\Omega_0 + \Omega_1(\pi_t q_t)^{\Omega_2}]^{m_1} + (1-\omega)\pi_t q_t - G_t \leqslant 0 \qquad (2.23)$$

(b) The second constraint which is a terminal constraint, i.e.

$$Q_T \geqslant 0$$

can be alternatively expressed as

$$\left(-\sum_{t-1}^{T} q_t + Q_0\right) \geqslant 0 \qquad (2.24)$$

by substituting in equation 2.5 of the macro model, where Q_0 is the initial value of the oil reserves which is given a priori.

2.4 The Optimal Control Problem

Expression 2.18 states the objective function as the sum of domestic and overseas investment, i.e.

$$\sum_{t=1}^{T} I_t + \sum_{t=1}^{T} B_t$$

In the light of the fact that the optimal control problem has to be stated in terms of state, control and exogenous variables, we need to replace the term B_t in the objective function with these variables.

We know from equation 2.9 of the macro model that

$$E_t = B_t + \frac{E_{t-1}}{1+h_t}$$

and from equation 2.20 which is the reduced form for the second state variable, we have,

$$E_t = -I_t + \upsilon_0(D_t + F_t) + \pi_t q_t + \frac{E_{t-1}}{1+h_t} - G_t$$

Therefore,

$$B_t = -I_t + \upsilon_0(D_t + F_t) + \pi_t q_t - G_t \tag{2.25}$$

Substituting equation 2.25 into the objective function, the optimal control problem can be restated as:

$$\text{Max} \sum_{t=1}^{T} I_t + \sum_{t=1}^{T} [-I_t + \upsilon_0(D_t + F_t) + \pi_t q_t - G_t]$$

subject to

$$\upsilon_2 D_t + \upsilon_1 F_t - I_t - m_0[\Omega_0 + \Omega_1(\pi_t q_t)^{\Omega_2}]^{m_1} + (1-\omega)\pi_t q_t - G_t \leqslant 0 \tag{2.26}$$

and

$$\left(-\sum_{t=1}^{T} q_t + Q_0\right) \geqslant 0$$

where

$$I_0 \geqslant 0 \qquad q_t \geqslant 0 \qquad \text{for } t = 1, \ldots, T$$

Setting the Lagrangian, W, we introduce λ_{1t} as the Lagrange multiplier associated with the first constraint in time period t and λ_2 as the Lagrange multiplier associated with the second constraint. Note that the second Lagrange multiplier does not have time as its subscript because λ_2 is related to only one time period, which is the terminal period.

$$
\begin{aligned}
W = \sum_{t=1}^{T} \{I_t + [-I_t + \upsilon_0(D_t + F_t) + \pi_t q_t - G_t] \\
+ \sum_{t=1}^{T} \lambda_{1t}\{-\upsilon_2 D_t - \upsilon_1 F_t + I_t \\
+ m_0[\Omega_0 + \Omega_1(\pi_{t-1}q_{t-1})^{\Omega_2}]^{m_1} \\
- (1-\omega)\pi_t q_t + G_t\} + \lambda_2\left(-\sum_{t=1}^{T} q_t + Q_0\right) \qquad (2.27)
\end{aligned}
$$

To obtain solutions to the above dynamic non-linear system we need to derive the Kuhn-Tucker[7] conditions. For the facility of exposition of these conditions and presenting the solutions in a more manageable way, it would be convenient to work with a more compact form of the above Lagrangian[8]. This would be achieved by substituting for I_t the expression K_t - K_{t-1}, using equation 2.7 of the macro-model. Now the variable K_t would act as both the state and control variables. The reduced form of the Lagrange multiplier takes the form:

$$
\begin{aligned}
W = \sum_{t=1}^{T} [\upsilon_0(D_t + F_t) + \pi_t q_t - G_t] \\
+ \sum_{t=1}^{T} \lambda_{1t}\{-\upsilon_2 D_t - \upsilon_1 F_t + K_t - K_{t-1} \\
+ m_0[\Omega_0 + \Omega_1(\pi_{t-1}q_{t-1})^{\Omega_2}]^{m_1} - (1-\omega)\pi_t q_t + G_t\} \\
+ \lambda_2\left(-\sum_{t=1}^{T} q_t + Q_0\right) \qquad (2.28)
\end{aligned}
$$

The Kuhn-Tucker conditions with respect to K_t for $t = 1,\ldots, T$ are as follows:

$$
\begin{aligned}
&\frac{\partial W}{\partial K_T} = \lambda_{1T} \leqslant 0 && K_T \geqslant 0 \\
&\frac{\partial W}{\partial K_{T-1}} = \upsilon_0\left(\frac{\partial D_T}{\partial K_{T-1}} + \frac{\partial F_T}{\partial K_{T-1}}\right) - \lambda_{1T} + \lambda_{1T-1} \\
&\qquad - \lambda_{1T}\left(\upsilon_2 \frac{\partial D_T}{\partial K_{T-1}} + \upsilon_1 \frac{\partial F_T}{\partial K_{T-1}}\right) \leqslant 0 && K_{T-1} \geqslant 0 \\
&\quad \vdots \qquad\qquad\qquad \vdots && (2.29) \\
&\frac{\partial W}{\partial K_{T-j}} = \upsilon_0\left(\frac{\partial D_{T-j+1}}{\partial K_{T-j}} + \frac{\partial F_{T-j+1}}{\partial K_{T-j}}\right) - \lambda_{1T-j+1} \\
&\qquad + \lambda_{1T-j} - \lambda_{1T-j+1}\left(\upsilon_2 \frac{\partial D_{T-j+1}}{\partial K_{T-j}} + \upsilon_1 \frac{\partial F_{T-j+1}}{\partial K_{T-j}}\right) \leqslant 0 \\
&\qquad \text{for } j = 1, \ldots, T-1 && K_{T-j} \geqslant 0
\end{aligned}
$$

The second control variable in the model for which we need to derive the Kuhn-Tucker condition is the volume of oil output in each period, q_t. The procedure is the same as above, where the derivatives with respect to K_t were determined. That is,

$$\frac{\partial W}{\partial q_T} = \pi_T + \lambda_{1T}[-(1-\omega)\pi_T] - \lambda_2 \leqslant 0 \qquad q_T \geqslant 0$$

$$\frac{\partial W}{\partial q_{T-1}} = \pi_{T-1} + \upsilon_0 \frac{\partial F_T}{\partial q_{T-1}} + \lambda_{1T}\left(-\upsilon_1 \frac{\partial F_T}{\partial q_{T-1}} + \frac{\partial V_T}{\partial q_{T-1}}\right)$$
$$+ \lambda_{1T-1}[-(1-\omega)\pi_{T-1}] - \lambda_2 \leqslant 0 \qquad q_{T-1} \geqslant 0$$

$$\frac{\partial W}{\partial q_{T-2}} = \pi_{T-2} + \upsilon_0 \frac{\partial F_T}{\partial q_{T-2}} + \upsilon_0 \frac{\partial F_{T-1}}{\partial q_{T-2}} + \lambda_{1T}\left(-\upsilon_1 \frac{\partial F_T}{\partial q_{T-2}}\right)$$
$$+ \lambda_{1T-1}\left(-\upsilon_1 \frac{\partial F_{T-1}}{\partial q_{T-2}} + \frac{\partial V_{T-1}}{\partial q_{T-2}}\right)$$
$$+ \lambda_{1T-2}[-(1-\omega)\pi_{T-2}] - \lambda_2 \leqslant 0 \qquad q_{T-2} \geqslant 0$$

$$\vdots \qquad (2.30)$$

$$\frac{\partial W}{\partial q_{T-j}} = \pi_{T-j} + \sum_{K=T-j+1}^{T} \upsilon_0 \frac{\partial F_K}{\partial q_{T-j}} + \sum_{K=T-j+1}^{T} \lambda_{1K}\left[-\upsilon_1 \frac{\partial F_K}{\partial q_{T-j}}\right]$$
$$+ \lambda_{1T-j+1} \frac{\partial V_{T-j+1}}{\partial q_{T-j}}$$
$$+ \lambda_{1T-j}[-(1-\omega)\pi_{T-j}] - \lambda_2 \leqslant 0 \qquad q_{T-j} \geqslant 0$$
$$j = 1, \ldots, T-1$$

In order to reach the solutions, we first concentrate on the derivative of the Lagrangian with respect to K_{T-j}. We know from equation 2.8 that F_t is a function of E_{t-1} and from the equation of motion 2.20 that E_t is a function K_t. Therefore,

$$\frac{\partial F_{T-j+1}}{\partial K_{T-j}} = \frac{\partial F_{T-j+1}}{\partial E_{T-j}} \cdot \frac{\partial E_{T-j}}{\partial K_{T-j}}$$

again from equation 2.20, and bearing in mind that $I_t = K_t - K_{t-1}$

$$\frac{\partial E_t}{\partial K_t} = -1$$

which in turn implies

$$\frac{\partial F_{T-j+1}}{\partial K_{T-j}} = -\frac{\partial F_{T-j+1}}{\partial E_{T-j}}.$$

As a result

$$\frac{\partial W}{\partial K_{T-j}} = \upsilon_0\left(\frac{\partial D_{T-j+1}}{\partial K_{T-j}} - \frac{\partial F_{T-j+1}}{\partial E_{T-j}}\right) - \lambda_{1T-j+1} + \lambda_{1T-j}$$
$$- \lambda_{1T-j+1}\left(\upsilon_2\frac{\partial D_{T-j+1}}{\partial K_{T-j}} - \upsilon_1\frac{\partial F_{T-j+1}}{\partial E_{T-j}}\right) \leqslant 0 \qquad K_{T-j} \geqslant 0$$

For the purpose of simplicity, we define

$$\frac{\partial D_{T-j+1}}{\partial K_{T-j}} \equiv D'_{T-j+1} \qquad \frac{\partial F_{T-j+1}}{\partial E_{T-j}} \equiv F'_{T-j+1} \tag{2.31}$$

for $j = 1, \ldots, T-1$.

By the complementary slackness conditions[9] for all the strictly positive values of K_t, the derivative $\partial W/\partial K_{T-j}$ must be equal to zero

$$\upsilon_0(D'_{T-j+1} - F'_{T-j+1}) - \lambda_{1T-j+1} + \lambda_{1T-j}$$
$$- \lambda_{1T-j+1}(\upsilon_2 D'_{T-j+1} - \upsilon_1 F'_{T-j+1}) = 0$$

There are obvious interdependencies between the Lagrangian multipliers λ_{1T-j} and λ_{1T-j+1}. We use this relationship to analyse the solutions. These are summarized below under five alternative conditions.

1 When both

$$\begin{aligned} \lambda_{1T-j} = \lambda_{1T-j+1} = 0, \\ \upsilon_0(D'_{T-j+1} - F'_{T-j+1}) = 0 \end{aligned} \tag{2.32}$$

which means

$$D'_{T-j+1} = F'_{T-j+1} \tag{2.33}$$

In other words when the balance of payments constraint is non-binding in both periods t and $t-1$, the optimal policy would be to equalize rates of return on domestic and foreign investment at the margin, that is, to invest in the sector with the highest rate of return.

2 When

$$\lambda_{1T-j} = 0$$

and

$$\lambda_{1T-j+1} > 0$$

then

$$\upsilon_0(D'_{T-j+1} - F'_{T-j+1}) - \lambda_{1T-j+1} - \lambda_{1T-j+1}(\upsilon_2 D'_{T-j+1} - \upsilon_1 F'_{T-j+1}) = 0$$

Rearranging terms gives

$$-\lambda_{1T-j+1}-\lambda_{1T-j+1}(\upsilon_2 D'_{T-j+1}-\upsilon_1 F'_{T-j+1}) = -\upsilon_0(D'_{T-j+1}-F'_{T-j+1})$$

Multiplying through a minus sign, and rearranging, we have

$$\lambda_{1T-j+1}[1+(\upsilon_2 D'_{T-j+1}-\upsilon_1 F'_{T-j+1})]=\upsilon_0(D'_{T-j+1}-F'_{T-j+1})$$

$$\lambda_{1T-j+1}=\frac{\upsilon_0(D'_{T-j+1}-F'_{T-j+1})}{[1+(\upsilon_2 D'_{T-j+1}-\upsilon_1 F'_{T-j+1})]}$$

Concentrating on the denominator and using the definition (2.22),

$$\upsilon_2 D'_{T-j+1}-\upsilon_1 F'_{T-j+1}=\upsilon_1 D'_{T-j+1}+\phi\xi D'_{T-j+1}-\upsilon_1 F'_{T-j+1}$$

$$\lambda_{1T-j+1}=\frac{\upsilon_0(D'_{T-j+1}-F'_{T-j+1})}{[1+\phi\xi D'_{T-j+1}+\upsilon_1(D'_{T-j+1}-F'_{T-j+1})]} \qquad (2.34)$$

The denominator of (2.34) can be shown to be positive. Hence, for λ_{1T-j+1} to be positive, the numerator of (2.34) has to be strictly positive, i.e.

$$\upsilon_0(D'_{T-j+1}-F'_{T-j+1})>0$$

since

$$\upsilon_0>0 \quad \text{then} \quad D'_{T-j+1}>F'_{T-j+1} \qquad (2.35)$$

3 When both

$$\lambda_{1T-j}>0$$

and

$$\lambda_{1T-j+1}>0$$

the relationship is indeterminate
i.e.

$$D'_{T-j+1}\gtreqless F'_{T-j+1} \qquad (2.36)$$

4 When

$$\lambda_{1T-j}>0$$

$$\lambda_{1T-j+1}=0$$

From equation 2.32 above we have:

$$\upsilon_0(D'_{T-j+1}-F'_{T-j+1})+\lambda_{1T-j}=0$$

which implies

$$\upsilon_0(F'_{T-j+1}-D'_{T-j+1})=\lambda_{1T-j} \qquad (2.37)$$

Since the right hand side is positive, for the left hand side to be positive

$$F'_{T-j+1} > D'_{T-j+1}$$

i.e. when the balance of payment constraint becomes non-binding a higher rate of return on overseas than home investment is required.

5 When we return to

$$\lambda_{1T-j} = 0$$

$$\lambda_{1T-j+1} = 0$$

again

$$F'_{T-j+1} = D'_{T-j+1}$$

meaning equalizing rates of return at the margin on home and overseas investment is optimal.[10]

To complete our solutions, we also have to analyse the Kuhn–Tucker conditions with respect to q_{T-j}, as shown above in (2.30). Concentrating on the general case, $\frac{\partial W}{\partial q_{T-j}}$ for $j = 1, \ldots, T-1$, by the complementary slackness condition, for $q_{T-j} > 0$, $\frac{\partial W}{\partial q_{T-j}} = 0$. This implies:

$$\pi_{T-j} + \sum_{K=T-j+1}^{T} (\upsilon_0 - \lambda_{1K}\upsilon_1)\frac{\partial F_K}{\partial q_{T-j}} = \lambda_2 + \lambda_{1T-j}[(1-\omega)\pi_{T-j}] - \lambda_{1T-j+1}\frac{\partial V_{T-j+1}}{\partial q_{T-j}} \tag{2.38}$$

$j = 1, \ldots, T-1$

Here again we need to analyse the interdependencies between the Lagrange multipliers. These are summarized below:

1 Consider the case when the balance of payments constraints are non-binding in all the K periods, i.e. when $\lambda_{1K} = 0$ for $K = T-j+1, \ldots, T$. Equation 2.38 becomes

$$\pi_{T-j} + \sum_{K=T-j+1}^{T} \upsilon_0 \frac{\partial F_K}{\partial q_{T-j}} = \lambda_2 \tag{2.38a}$$

where $j = 1, \ldots, T-1$

and the equilibrium solution is the following:

$$\frac{\Delta\pi_{T-j+1}}{\pi_{T-j}} = \frac{\upsilon_0(1+\upsilon_0\theta)^{j-1}}{1+\sum_{k=T-j+1}^{T-1} \upsilon_0\,(\partial F_k/\partial E_{T-j})} \tag{2.39}$$

The derivation of this solution, which is lengthy and involves a large number of steps, is given in the Appendix. Since the expression inside the square brackets is positive—this is also proved in the Appendix, – we can conclude that

$$\frac{\Delta\pi_{T-j+1}}{\pi_{T-j}} > \theta \tag{2.40}$$

This solution has important implications as far as the optimal depletion of the oil stock is concerned. It implies that when the balance of payments constraint is non-binding the rate of change of the resource prices has to be greater than the rate of return on investment. It is important to note that the solution reached here is a modified version of Hotelling's solution.[11] For Hotelling's rule to hold, the proportional rate of change of prices must equal the going rate of interest. This solution is reached where the miner operates in a competitive market and acts primarily as the seller of the resource. In our solution where the state is the owner of the resource, (i.e. both producer and consumer of the resource), the optimal depletion calls for the rate of return to be less than the percentage rate of price change, which in turn calls for a slower depletion rate for oil. We shall further elaborate on this solution later in section 2.5

2 The second condition is when

$$\lambda_{1T-j} = 0$$

$$\lambda_{1T-j+1} > 0$$

and

$$\lambda_{1k} = 0 \qquad k = T-j+2, \dots, T$$

For strictly positive values of q_{T-j} we have

$$\pi_{T-j} + \sum_{K=T-j+1}^{T} (\upsilon_0 - \lambda_{1K}\upsilon_1)\frac{\partial F_k}{\partial q_{T-j}} + \lambda_{1T-j+1}\frac{\partial V_{T-j+1}}{\partial q_{T-j}} - \lambda_2 = 0$$

since $\lambda_{1K} = 0$ for $K = T-j+2, \dots, T$ the above expression becomes:

$$\pi_{T-j} + \sum_{K=T-j+1}^{T} \upsilon_0 \frac{\partial F_k}{\partial q_{T-j}} - \lambda_{1T-j+1}\upsilon_1 \frac{\partial F_{T-j+1}}{\partial q_{T-j}} + \lambda_{1T-j+1} \frac{\partial V_{T-j+1}}{\partial q_{T-j}} - \lambda_2 = 0$$

$$\lambda_{1T-j+1}\left(\upsilon_1 \frac{\partial F_{T-j+1}}{\partial q_{T-j}} - \frac{\partial V_{T-j+1}}{\partial q_{T-j}}\right) = \pi_{T-j} + \sum_{k=T-j+1}^{T} \upsilon_0 \frac{\partial F_{T-j+1}}{\partial q_{T-j}} - \lambda_2$$

$$\lambda_{1T-j+1} = \frac{\pi_{T-j} + \sum_{k=T-j+1}^{T} \upsilon_0(\partial F_k / \partial q_{T-j}) - \lambda_2}{\upsilon_1(\partial F_{T-j+1}/\partial q_{T-j}) - (\partial V_{T-j+1}/\partial q_{T-j})} \tag{2.41}$$

The numerator of the above can be shown to be positive.

$$\frac{\partial F_{T-j+1}}{\partial q_{T-j}} = \frac{\partial F_{T-j+1}}{\partial E_{T-j}} \cdot \frac{\partial E_{T-j}}{\partial q_{T-j}} = \theta \pi_{T-j} > 0$$

$$\frac{\partial V_{T-j+1}}{\partial q_{T-j}} = \frac{\partial V_{T-j+1}}{\partial S_{T-j}} \cdot \frac{\partial S_{T-j}}{\partial q_{T-j}}$$

where

$$\frac{\partial V_{T-j+1}}{\partial S_{T-j}} < 0 \quad \text{and} \quad \frac{\partial S_{T-j}}{\partial q_{T-j}} > 0$$

Thus,

$$\left[-\frac{\partial V_{T-j+1}}{\partial q_{T-j}}\right] > 0$$

Since the denominator of equation 2.41 is positive, for λ_{1T-j+1} to be positive the numerator of the right hand side has to be positive.

i.e.,

$$\pi_{T-j} + \sum_{K=T-j+1}^{T} \upsilon_0 \frac{\partial F_K}{\partial q_{T-j}} > \lambda_2 \tag{2.42}$$

3 When $\lambda_{1T-j} > 0$, $\lambda_{1T-j+1} > 0$ and $\lambda_{1K} = 0$ for $K = T-j+2, \ldots, T$, we obtain a modified version of equation 2.38 above:

$$\pi_{T-j} + \sum_{K=T-j+1}^{T} \upsilon_0 \frac{\partial F_K}{\partial q_{T-j}} - \lambda_{1T-j+1}\upsilon_1 \frac{\partial F_{T-j+1}}{\partial q_{T-j}} = \lambda_2 + \lambda_{1T-j}[(1-\omega)\pi_{T-j}] - \lambda_{1T-j+1} \frac{\partial V_{T-j+1}}{\partial q_{T-j}}$$

Since each of the terms on the right hand side are positive, the left hand side of the above equation has to be positive.

i.e.

$$\pi_{T-j} + \sum_{K=T-j+1}^{T} v_0 \frac{\partial F_K}{\partial q_{T-j}} > \lambda_{1T-j+1} v_1 \frac{\partial F_{T-j+1}}{\partial q_{T-j}}$$

Therefore, without specific values of the parameters, it would be difficult to obtain any more specific solution.

4 Alternatively when

$$\lambda_{1T-j} > 0$$

$$\lambda_{1T-j+1} = 0$$

and

$$\lambda_{1K} = 0 \qquad \text{for} \qquad K = T-j+2, \ldots, T$$

$$\pi_{T-j} + \sum_{K=T-j+1}^{T} v_0 \frac{\partial F_K}{\partial q_{T-j}} = \lambda_2 + \lambda_{1T-j}[(1-\omega)\pi_{T-j}]$$

Rearranging

$$\lambda_{1T-j} = \frac{\pi_{T-j} + \sum_{K=T-j+1}^{T} v_0 (\partial F_K / \partial q_{T-j}) - \lambda_2}{(1-\omega)\pi_{T-j}}$$

Since $0 < \omega < 1$, the denominator must be positive. Thus for $\lambda_{1T-j} > 0$, the numerator must be positive too, which implies again

$$\pi_{T-j} + \sum_{K=T-j+1}^{T} v_0 \frac{\partial F_K}{\partial q_{T-j}} > \lambda_2$$

This is the same as the solution reached under condition 2.

5 This last stage is when the constraints are non-binding over all the remaining periods.

$$\lambda_{1K} = 0 \qquad K = T-j, \ldots, T$$

This is the same as Condition 1 outlined earlier and the same solutions hold.

The economic implication of conditions 2 and 4 when the constraint is binding in periods *T-j* and *T-j*+1 is the following: when the balance of payments constraint becomes binding, the optimal policy would be to cut down on oil production. The rationale for this solution is that if the constraints were not binding, the inequality 2.42 would become an equality such that

$$\pi_{T-j} + \sum_{K=T-j+1}^{T} v_0 \frac{\partial F_K}{\partial q_{T-j}} = \lambda_2 \qquad (2.43)$$

One has to investigate what causes the balance of payments constraint to become binding and to turn the equality 2.43 into an inequality. Clearly one of the terms on the left hand side of equation 2.43 has to rise to change it into the inequality 2.42. Of the two variables, either the price of oil π_{T-j} has to increase or the term $\partial F_K/\partial q_{T-j}$ has to rise. Bearing in mind that $\partial F_K/\partial q_{T-j}$ is itself a function of the prices of oil over time it implies that petroleum prices π_{T-j}'s are higher in inequality 2.42 compared with equality 2.43. This, in turn, suggests that higher petroleum prices can make the balance of payment constraint binding. The reason for this is the strengthening effect on sterling that higher oil revenue can cause (as shown in equation 2.17 of the macro model). The high exchange rate in turn forces the non-oil exports V_t to decline and deteriorate the trade balance. Under these circumstances, since the oil price is exogenous the only variable we can control is the value of q_t, the production of oil, and a reduction of q_t would make the constraint non-binding. Thus, as far as policy is concerned, this implies that when the balance of payments constraint becomes binding, the optimal policy would be to cut back on oil production.

2.5 Summary Of The Analytical Solutions

The implications of the solutions stated in section 2.4 are considered briefly here. The interrelationship of the Lagrange multipliers were analysed separately for the variable K_t (domestic non-oil capital formation) and q_t (the production of oil) under five different conditions, or stages. However, since the different stages related to the two variables coincide with one another—for instance, stage 1, related to the variable K_t, is the same as stage 1 related to variable q_t—their implications can be explained simultaneously. As far as policy is concerned these concurrent stages can be portrayed as described next.

During stage 1 when the balance of payments constraint is non-binding over a certain period of time, equalizing rates of return at the margin between home and overseas investment is the optimal policy. At the same time, under these conditions the rule for the production of oil should be according to a modified version of Hotelling's rule. The optimal depletion rate calls for the rate of change of oil prices over time to be less than the rate of interest. It is less by the value of the fraction in equation 2.39 consisting of υ_0, a combination of some parameters, and the yield on overseas investment. υ_0, which is defined in equation 2.19, comprises the parameters average propensity to consume, income-related tax coefficient and expenditure tax coefficient. Effectively, v_0 is some form of savings propensity which takes into account tax leakages. So depending on the magnitude of these parameters, the rate of return on overseas investment and the rate of change of the price of oil, the depletion of petroleum is determined.

When the constraint first becomes binding, i.e. when $\lambda_{1T-j}=0$ and $\lambda_{1T-j+1}>0$ (stage 2), this implies that imports have to be kept at their minimum level. The optimal solution says invest overseas and cut back on oil production. This is because, given the structure of the economy, if under these circumstances investment at home takes place, more imports will be required. The higher volume of imports would not be offset by the non-oil exports alone, due to the high exchange rate, and most of the oil earnings of foreign exchange would have to be used up to pay for imports. Moreover, the economy would be put on an expansion path such that when the oil runs out and the foreign exchange dwindles, a gap would emerge between the receipts and the payments.

When the balance of payments constraint is binding over consecutive periods, i.e. when $\lambda_{1T-j}>0$ and $\lambda_{1T-j+1}>0$, we have labelled this as stage 3. As far as the allocation of investment between home and overseas is concerned, the solution is indeterminate and one has to resort to stage 4 which gives a clear-cut direction for policy. However, as far as the depletion rule is concerned, we obtain a conclusive solution: namely, cut back on oil extraction. In other words when the balance of payments is binding over a prolonged period, the optimal policy is to reduce petroleum production, and this will, in turn, reduce the exchange rate and help improve non-oil exports.

Stage 4 is when the constraint is binding in period $T-j$ and non-binding in $T-j+1$. Then, the optimal policy calls for more investment at home, albeit at a lower rate of return than that achieved overseas, accompanied by a reduction in petroleum extraction. The rationale for domestic investment is to boost non-oil exports and to gain other sources of foreign exchange besides oil. In other words, at this stage the economy has to plan to reduce its dependence on petroleum. Although import requirements will increase as a result of the high domestic investment, as was pointed out in stage 2, by this time the income from overseas assets would have sufficiently grown (having accummulated during stages 2 and 3) to offset for the dwindling oil revenue. Investment in non-oil exports at home during stage 4 will ensure a smooth transition into a state with a non-replenishable source of income.

Appendix

The Kuhn-Tucker conditions for the derivatives of the Lagrangian with respect to q_{T-j} for j=1,...,$T-1$ are analysed in the text. The solutions are outlined under five different alternative conditions when the balance of payments constraints are non-binding over K periods, i.e., $\lambda_{1K}=0$ for $K=T-j+1,...,T$.

If all the λ_{1K}'s are set equal to zero, we have equation 2.38a.

$$\pi_{T-j+1} + \sum_{K=T-j+1}^{T} \upsilon_0 \frac{\partial F_K}{\partial q_{T-j}} = \lambda_2$$

which means for all the $T-j$ periods (where j=1,...,$T-1$) the left-hand side of equation is equal to the constant λ_2. For instance,

$$\pi_{T-j+1} + \sum_{K=T-j+2}^{T} \upsilon_0 \frac{\partial F_K}{\partial q_{T-j+2}} = \lambda_2$$

This implies the left hand side of the above two equations are equal, i.e.,

$$\pi_{T-j} + \sum_{K=T-j+1}^{T} \upsilon_0 \frac{\partial F_K}{\partial q_{T-j}} = \pi_{T-j+1} + \sum_{K=T-j+2}^{T} \upsilon_0 \frac{\partial F_K}{\partial q_{T-j+2}}$$

Multiplying through a minus sign and rearranging gives:

$$-\pi_{T-j} + \pi_{T-}j_{+1} - \sum_{K=T-j+1}^{T} \upsilon_0 \frac{\partial F_K}{\partial q_{T-j}} + \sum_{K=T-j+2}^{T} \upsilon_0 \frac{\partial F_K}{\partial q_{T-j+1}} = 0$$

$$\Delta\pi_{T-j+1} - \sum_{K-T-j+1}^{T} \upsilon_0 \frac{\partial F_K}{\partial q_{T-j}} + \sum_{K=T-j+2}^{T} \upsilon_0 \frac{\partial F_K}{\partial q_{T-j+1}} = 0 \tag{A2.1}$$

Bearing in mind that from equation 2.8 of the macromodel,

$$F_t = \theta E_{t-1}$$

from the equation of motion 2.20

$$E_t = - I_t + \upsilon_0 D_t + \upsilon_0 \theta E_{t-1} + \pi_t q_t + E_{t-1} - G_t$$

Stating the derivatives of F_K with respect to q_{T-j} for $K=T-j+1,\ldots,T$

$$\frac{\partial F_{T-j+1}}{\partial q_{T-j}} = \frac{\partial F_{T-j+1}}{\partial E_{T-j}} \cdot \frac{\partial E_{T-j}}{\partial q_{T-j}} = \theta\pi_{T-j}$$

$$\frac{\partial F_{T-j+2}}{\partial q_{T-j}} = \frac{\partial F_{T-j+2}}{\partial E_{T-j+1}} \cdot \frac{\partial E_{T-j+1}}{\partial E_{T-j}} \cdot \frac{\partial E_{T-j}}{\partial q_{T-j}} = \theta(1+\upsilon_0\theta)\pi_{T-j}$$

$$\vdots \tag{A2.2}$$

$$\frac{\partial F_{T-1}}{\partial q_{T-j}} = \frac{\partial F_{T-1}}{\partial E_{T-2}} \cdot \frac{\partial E_{T-2}}{\partial E_{T-3}} \cdots \frac{\partial E_{T-j}}{\partial q_{T-j}} = \theta(1+\upsilon_0\theta)^{j-2}\pi_{T-j}$$

$$\frac{\partial F_{T}}{\partial q_{T-j}} = \frac{\partial F_{T}}{\partial E_{T-1}} \cdots \frac{\partial E_{T-j}}{\partial q_{T-j}} = \theta(1+\upsilon_0\theta)^{j-1}\pi_{T-j}$$

Similarly, stating the derivatives of F_K with respect to q_{T-j+1} when $K = T-j+2, \ldots, T$,

$$\frac{\partial F_{T-j+2}}{\partial q_{T-j+1}} = \frac{\partial F_{T-j+2}}{\partial E_{T-j+1}} \cdot \frac{\partial E_{T-j+1}}{\partial q_{T-j+1}} = \theta\pi_{T-j+1}$$

$$\frac{\partial F_{T-j+3}}{\partial q_{T-j+1}} = \frac{\partial F_{T-j+3}}{\partial E_{T-j+2}} \cdot \frac{\partial E_{T-j+2}}{\partial E_{T-j+1}} = \theta(1+\upsilon_0\theta)\pi_{T-j+1}$$

$$\vdots \tag{A2.3}$$

$$\frac{\partial F_{T}}{\partial q_{T-j+1}} = \frac{\partial F_{T}}{\partial E_{T-1}} \cdot \frac{\partial E_{T-1}}{\partial E_{T-2}} \cdots \frac{\partial E_{T-j+1}}{\partial q_{T-j+1}} = \theta(1+\upsilon_0\theta)^{j-2}\pi_{T-j+1}$$

It can be clearly seen that the outcome of the first derivative of the set of derivatives A 2.3 corresponds with the first one in A 2.2. That is to say:

$$\frac{\partial F_{T-j+1}}{\partial q_{T-j}} = \theta\pi_{T-j}$$

and

$$\frac{\partial F_{T-j+2}}{\partial q_{T-j+1}} = \theta\pi_{T-j+1}$$

It can be shown that[1]

$$\pi_{T-j+1} = \pi_{T-j}\left(1+\frac{\Delta\pi_{T-j+1}}{\pi_{T-j}}\right) \tag{A2.4}$$

[1]The steps are the following:
If we add and subtract π_{T-j} to π_{T-j+1} the value remains unchanged:

$$\pi_{T-j+1} = \pi_{T-j} + \pi_{T-j+1} - \pi_{T-j} = \pi_{T-j} + \Delta\pi_{T-j+1}$$

If we multiply and divide the right hand side by π_{T-j} the expression remains unchanged:

$$\pi_{T-j+1} = (\pi_{T-j} + \Delta\pi_{T-j+1})\left(\frac{\pi_{T-j}}{\pi_{T-j}}\right) = \pi_{T-j}\left[1+\frac{\Delta\pi_{T-j+1}}{\pi_{T-j}}\right]$$

Substituting the above expression into the derivatives (40), they become

$$\frac{\partial F_{T-j+2}}{\partial q_{T-j+1}} = \theta \pi_{T-j}\left(1+\frac{\Delta \pi_{T-j+1}}{\pi_{T-j}}\right)$$

$$\frac{\partial F_{T-j+3}}{\partial q_{T-j+1}} = \theta(1+\upsilon_0\theta)\,\pi_{T-j}\left(1+\frac{\Delta \pi_{T-j+1}}{\pi_{T-j}}\right)$$

$$\vdots \qquad \text{(A2.5)}$$

$$\frac{\partial F_{T}}{\partial q_{T-j+1}} = \theta(1+\upsilon_0\theta)^{j-2}\,\pi_{T-j}\left(1+\frac{\Delta \pi_{T-j+1}}{\pi_{T-j}}\right)$$

Notice in the set of derivatives A 2.5 there are $j-2$ terms and in the set of derivatives A 2.2 there are $j-1$ terms. The last derivative in the latter set, $\partial F_T/\partial q_{T-j}$ does not have a counterpart in the set A 2.5.

If we separate the summation in A2.1 $\sum_{K=T-j+1}^{T} \upsilon_0(\partial F_K/\partial q_{T-j})$ into two separate components: a summation going from $T-j+1$ to $T-1$ and one term for time period T, we have

$$\Delta \pi_{T-j+1} - \sum_{K=T-j+1}^{T-1} \upsilon_0 \frac{\partial F_K}{\partial q_{T-j}} - \upsilon_0 \frac{\partial F_T}{\partial q_{T-j}} + \sum_{K=T-j+2}^{T} \upsilon_0 \frac{\partial F_K}{\partial q_{T-j+1}} = 0 \qquad \text{(A2.6)}$$

Substituting the relationship 2.42 above

$$\frac{\partial F_{T-j+2}}{\partial q_{T-j+1}} = \frac{\partial F_{T-j+1}}{\partial E_{T-j}}\,\pi_{T-j}\left(1+\frac{\Delta \pi_{T-j+1}}{\pi_{T-j}}\right)$$

$$\vdots$$

$$\frac{\partial F_{T}}{\partial q_{T-j+1}} = \frac{\partial F_{T-1}}{\partial E_{T-j}}\,\pi_{T-j}\left(1+\frac{\Delta \pi_{T-j+1}}{\pi_{T-j}}\right)$$

which implies:

$$\sum_{K=T-j+2}^{T} \upsilon_0 \frac{\partial F_K}{\partial q_{T-j+1}} = \sum_{K=t-j+1}^{T-1} \upsilon_0 \frac{\partial F_K}{\partial E_{T-j}}\,\pi_{T-j}\left(1+\frac{\Delta \pi_{T-j+1}}{\pi_{T-j}}\right) \qquad \text{(A2.7)}$$

and bearing in mind

$$\frac{\partial F_K}{\partial q_{T-j}} = \frac{\partial F_K}{\partial E_{T-j}} \cdot \pi_{T-j} \qquad \text{for } K = 1, \ldots, T.$$

Substituting the above relationship and equation A2.7 into A2.6 we have:

$$\Delta\pi_{T-j+1} - \sum_{K=T-j+1}^{T-1} \upsilon_0 \frac{\partial F_K}{\partial E_{T-j}} \pi_{T-j} - \upsilon_0 \frac{\partial F_T}{\partial E_{T-j}} \cdot \pi_{T-j} + \sum_{K=T-j+1}^{T-1} \upsilon_0 \frac{\partial F_K}{\partial E_{T-j}} \pi_{T-j}\left(1 + \frac{\Delta\pi_{T-j+1}}{\pi_{T-j}}\right) = 0 \quad \text{(A2.8)}$$

Cancelling identical terms the expression reduces to:

$$\Delta\pi_{T-j+1} - \upsilon_0 \frac{\partial F_T}{\partial E_{T-j}} \pi_{T-j} + \sum_{K=T-j+1}^{T-1} \upsilon_0 \frac{\partial F_K}{\partial E_{T-j}} \pi_{T-j} \frac{\Delta\pi_{T-j+1}}{\pi_{T-j}} = 0 \quad \text{(A2.9)}$$

Substituting for $(\partial F_T / \partial E_{T-j}) = \theta(1+\upsilon_0\theta)^{j-1}\pi_{T-j}$ from the derivative (A2.2) above, (A2.9) becomes

$$\Delta\pi_{T-j+1} - \upsilon_0[\theta(1+\upsilon_0\theta)^{j-1}]\pi_{T-j} + \sum_{K=T-j+1}^{T-1} \upsilon_0 \frac{\partial F_K}{\partial E_{T-j}} \pi_{T-j} \frac{\Delta\pi_{T-j+1}}{\pi_{T-j}} = 0$$

Dividing through by π_{T-j}

$$\frac{\Delta\pi_{T-j+1}}{\pi_{T-j}} - \upsilon_0[\theta(1+\upsilon_0\theta)^{j-1}]\frac{\pi_{T-j}}{\pi_{T-j}} + \sum_{K=T-j+1}^{T-1} \upsilon_0 \frac{\partial F_K}{\partial E_{T-j}} \frac{\Delta\pi_{T-j+1}}{\pi_{T-j}} = 0$$

Rearranging

$$\frac{\Delta\pi_{T-j+1}}{\pi_{T-j}}\left(1 + \sum_{K=T-j+1}^{T-1} \upsilon_0 \frac{\partial F_K}{\partial E_{T-j}}\right) - \upsilon_0[\theta(1+\upsilon_0\theta)^{j-1}] = 0$$

$$\therefore \quad \frac{\Delta\pi_{T-j+1}}{\pi_{T-j}} = \left[\frac{\upsilon_0(1+\upsilon_0\theta)^{j-1}}{1 + \sum_{K=T-j+1}^{T-1} \upsilon_0 \frac{\partial F_K}{\partial E_{T-j}}}\right]\theta \quad \text{(A2.10)}$$

where $j = 2, \ldots, T-1$

Equation A2.10 is the solution that one obtains when the balance of payments constraint is non-binding over the planning period. The expression inside the square bracket in A2.10 can be shown to be positive and less than 1.

$$0 < \left[\frac{\upsilon_0(1+\upsilon_0\theta)^{j-1}}{1 + \sum_{K=T-j+1}^{T-1} \upsilon_0 \frac{\partial F_K}{\partial E_{T-j}}}\right] < 1 \quad \text{(A2.11)}$$

$j = 2, \ldots, T-1$

We first show that υ_0 is positive and less than one. υ_0 is defined in equation 2.19 of text as

$$\begin{aligned}\upsilon_0 &= 1+\epsilon\eta+\sigma\eta-\epsilon\sigma\eta-\eta\\ &= 1+\eta(\epsilon+\sigma-\epsilon\sigma-1)\end{aligned}$$

But by assumption

$$0<\eta<1$$

$$0<\epsilon<1$$

$$u<\sigma<1$$

Hence

$$0<\epsilon<\frac{1-\sigma}{1-\sigma}$$

$$0<\epsilon(1-\sigma)<1-\sigma$$

$$0<\epsilon+\sigma-\epsilon\sigma<1$$

$$-1<\epsilon+\sigma-\epsilon\sigma-1<0$$

$$-1<\eta(\epsilon+\sigma-\epsilon\sigma-1)<0$$

and, therefore,

$$0<\upsilon_0<1$$

The value of θ, the real rate of return on overseas investment, is shown to be positive and less than one in the specification of the model.

So far we have shown that the numerator of A2.11 is strictly positive. We now have to show that the denominator is strictly positive and less than the numerator. This becomes more apparent when we expand the summation term in the denominator working through from the last period

$$\frac{\partial F_{T-1}}{\partial E_{T-j}} = \theta(1+\upsilon_0\theta)^{j-2}$$

$$\frac{\partial F_{T-2}}{\partial E_{T-j}} = \theta(1+\upsilon_0\theta)^{j-3} \qquad \text{(A2.12)}$$

$$\vdots$$

$$\frac{\partial F_{T-j+1}}{\partial E_{T-j}} = \theta$$

Summing all the terms in A2.12 and substituting it in the denominator of A2.11, we have

$$\left[\frac{\upsilon_0\theta(1+\upsilon_0\theta)^{j-1}}{1+\sum_{K=T-j+1}^{T-1}\upsilon_0[\theta(1+\upsilon_0\theta)^{j-2}+\theta(1+\upsilon_0\theta)^{j-3}+\cdots+\theta(1+\upsilon_0\theta)+\theta]}\right]$$

Having shown that both θ and υ_0 are positive, the denominator of the above expression is necessarily positive. Comparing the denominator of A2.13 with the numerator, it can easily be verified that the numerator is smaller than the denominator, for only one of the terms inside the summation is nearly as large as the numerator and we can assert

$$\upsilon_0\theta(1+\upsilon_0\theta)^{j-1}<1+\sum_{K=T-j+1}^{T-1}\upsilon_0[\theta(1+\upsilon_0\theta)^{j-2} +\cdots+\theta(1+\upsilon_0\theta)+\theta]$$

which proves the condition A2.11 above. This in turn makes the solution A2.10 hold.

3 Survey of The Principal UK Macroeconomic Models

3.1 Introduction

In this chapter we examine some of the major macroeconomic models of Britain and put forward a brief summary of each model. The purpose of this summary is twofold:

(i) to reveal that none of the existing models have dealt specifically with the problem set out in this study and, hence, to show the necessity to formulate our own macroeconomic model;

(ii) to provide the relevant information for when we come to quantify the parameters of our model in chapter 5. In other words, to provide a useful basis whenever there is a similarity between the specification of some of the equations of our model and those of the large econometric models.

Six main large econometric models of the UK are examined here. These are the models of Her Majesty's Treasury, the National Institute, the Cambridge Growth Project, the London Business School, the Bank of England and the Cambridge Policy Group. A brief outline of the theoretical structure of each model is given, and their distinguishing features and general aims and approaches are discussed.

In preparing this survey the models are primarily viewed with respect to their relevance to the question posed in this study. As it turns out, none of the existing large econometric models determine any depletion policies for the North Sea resource, although the oil sector appears as an exogenous variable in various equations in these models. This was true, at least, at the time this research was completed, it may well be that some of these models will be expanded to treat the question of depletion policies. But this factor has not so far been present in any of the published documents explaining the structure of such models.

Another important characteristic of these models is that they are all formulated essentially to cater for short-run problems, although some of them do put forward medium-term forecasts. In this respect they are not suitable for analysing long-term problems of the type posed in this book, ones that span over two and a half decades. Thus in the light of the above two factors it became essential to construct our own long-run macroeconomic model—this was specified in chapter 2.

3.2 A Brief Review of The Major UK Macroeconomic Models in Their Treatment of The Oil Sector

In model construction there have been basically two schools of thought. The more traditional approach has been to derive a formal functional form (based on economic theory) for the equations of the model, and then obtain the best parameter estimates. The other approach has been to make as few a priori restrictions as possible, to postulate no special theoretical function, but instead to allow estimation procedure to obtain the best possible fit. In practice, however, no model-builder adopts too extreme a position either way, and a compromise is usually reached.

In reviewing the following six models we shall point out which approach has been adopted for the construction and estimation of each model, and to what extent it has been a compromise between the two methods.

(i) *The HM Treasury Model*

The Treasury macroeconomic model is fairly large, even in comparison with the other large macromodels, with about 500 equations and well in excess of 700 variables. This does not, however, represent the full range or depth of the Treasury modelling endeavour; it is in fact only one of a group of interacting models. The other models in the group are the World Economic Prospects, Debt Interest, the Monetary Sector, Overseas Capital Flows and the Industrial models. Each individual model can be used to feed estimates into the main model, and this can, in turn, return other estimates to the sub-models. The system has the advantage of being able to reap the benefits of disaggregation while at the same time minimizing the costs of increased complexity. However, as the satellite models grow in complexity and the number of links with the main model increases, it becomes very much harder to separate the two. This situation has been increasingly true of the monetary sector model in particular, in which the links with the main model have proliferated to the point where meaningful separation of the two is difficult.

The intended purpose of the model is to provide short and medium-term forcasts and a simulation of the economy on a quarterly basis. Its basic approach is geared towards very effective short term modelling (three to four years), which is achieved by the widespread use of complex lag structures in the theoretical formalization of individual equations. This makes the model especially suitable for short-term use since quarterly fluctuations can be modelled very effectively.

The Treasury model is fundamentally a Keynesian one, quantities being determined solely on the demand side. The notion of the supply of output is generally not considered. Consumption in the short run is a function of current and lagged disposable income, lagged consumption, unemployment, personal wealth, the real interest rate and various dummy variables. Investment is a function of depreciation of the capital stock, manufacturing output lagged over 12 quarters, the internal cash flow of companies lagged over 7 quarters and interest rates lagged over 14 quarters.

The North Sea sector is not modelled as a whole, although it does appear as exogenous variables in other sectors, e.g. a determinant of imported oil. But so far (up till the time of this study) the model has not had a separate energy sector. Most important of all, the question of depletion of the North Sea oil resource is not even considered in the model, but is treated as given.

The model is generally estimated on a single equation basis using ordinary least squares. When it is used for simulation purposes the user is able to impose a fairly considerable degree of individual influence over the workings of the model by two means. The first is simply by choosing the values of the exogenous variables; the second is by specifying the size and sign of the residuals of the equations. Some areas of the model have not in fact been estimated but parameter values have been imposed judgmentally. Basically, this has occurred wherever statistical problems have prevented the estimation techniques from picking up firm a priori relationships. On the whole the model is too large and complex to allow a direct formal sequential solution to be obtained. As a result, it is solved by the use of a Gauss-Seidel iterative procedure.[1]

The Treasury model is characterized predominantly by two distinctive features. These are its size and its extensive use of lag structure in order to obtain a model that is sensitive down to a quarterly level. Both features are necessary and desirable in a model which is being used by a department such as the Treasury. The large size allows a detailed representation of alternative government policies, and the lag structure greatly improves the short-term properties of the model. It is, however, precisely these two properties which make it unsuitable to use for long-term planning, particularly for the type of problems set out in this study.

(ii) *The National Institute Model*

The National Institute Model has had a relatively long tradition of economic modelling dating back to 1969 when a small twenty-equation model was set up and used for basic forecasting. Between 1973 and 1974 the small model was revised and reformulated and a larger one with 75 equations emerged as the offspring of the first model. The limitations of this second, small model were quickly appreciated, however, particularly after the structural changes brought about by the oil crisis. During 1974–1975 a project was set up to construct an entirely new and much larger model, resulting in a considerably changed and enlarged model with some 120 equations.

During a set of simulation exercises, however, it was discovered that this model did not perform as well as the old one in certain respects. Consequently, the investment and pricing sectors of the old model were adapted and introduced into the new one. This improved the overall operation of the model in terms of statistical fit, but it also entailed a loss in theoretical consistency. It is this hybrid model that has formed the basis of the current National Institute model although it has undergone further modifications and re-estimations.

The basic structure of the model is well in line with the other large models. It has a Keynesian, income-expenditure structure where quantities are essentially demand-determined. Gross domestic product is determined by the components of aggregate demand: consumption, investment, and exports, with adjustment for factor cost. All are determined endogenously while government expenditure is projected as the main aggregate exogenous variable.

The National Institute model falls fairly consistently in the centre of the spectrum of the large models whichever aspect of the model is being examined. The Institute takes a fairly central position over the question of the a priori specification of equations. The functional forms used by the Institute are also generally fairly conventional; moderately complex lag structures are used in some of the equations, e.g. exports. The model is not over-large, and so is relatively easy to comprehend and to handle.

As far as the treatment of the oil sector is concerned, the model does not have a separate sector for this resource. Although research is carried out at the National Institute on various aspects of energy, and the information is fed into some of their equations, the model does not concentrate on the questions of the rate of oil depletion for the North Sea.

From the point of view of assigning parameters to the model developed in this study, the National Institute model is of little direct relevance because of its emphasis on short term, quarter-by-quarter adjustments and the fact that most of the equations exhibit more variables than the basic theory might suggest. However, the Institute model is theoretically well founded and small enough to provide an approximate check and comparison for the parameterization of our model.

(iii) *The Cambridge Growth Project Model*

The Cambridge Growth Project, which is also referred to as the Cambridge Multisectoral Domestic model, is now one of the longest-running and best-established macro-models of its sort in Britain, spanning over twenty year of continuous activity. Because the formal model has evolved gradually over a long period and it is not surprising that it is probably the most distinctive of the larger models. Even the stated objective of the model is different from that of most of the others in that it aims to provide projections over the medium (six years) to long term, rather than the short term. The Growth project also probably takes the most extreme view in the theory versus empirical estimation of equations, and opts for theoretical specification of the equations.

The model exists on two distinct levels: the fully-specified model of some 700 equations, with a high degree of disaggregation, and a smaller, condensed version, which in nineteen equations gives a concise picture of the model's structure. The condensed version can be given parameters derived directly from aggregation of the parameter values of the larger model. This process is possible, mainly, because the model is explicitly not a short-term model, which means that there is no need for complex quarterly lag structures. It is perhaps this factor which most clearly distinguishes the Growth Project from the other models.

In its broad outline, however, the Growth Project model does not differ greatly from the other large models. It has conventional income-expenditure structure where quantities are still essentially determined on the demand side of the market. Demand for total output is divided into two main sectors of about equal numerical size, the ordinary demand for final products and industry's demands for inputs to the immediate production process. The first sector is modelled along conventional lines by splitting it into consumption, investment, government and foreign demand. The second sector is shown by a detailed model of the industrial sector, using a modified Leontief[2] input-output formulation. This allows the model to represent the effects of major structural changes in the economy much more effectively. For example, if there is a major change in the relative price of oil, the model could reflect the likely changes that this might bring about within the industrial sector itself. A conventional macroeconomic model would not be able to reflect this change and would therefore be expected to do significantly worse.

Demand by private consumers is modelled on the basis of conventional consumption analysis, where consumption is disaggregated into thirty-eight non-durable sectors and four durable sectors. Government expenditure is generally treated as exogenous although in some simulations it can play the more active role of a policy instrument. The demand for investment is also built up from a large number of individual industry functions where each function has expected output and relative prices as its arguments. The final sector of demand, exports, is also disaggregated, and each equation is related to foreign output, relative prices and pressure of capacity. The sum of these sectors gives total output, which in turn gives the total demand for employment by means of an inverse production function. The Growth Project model as a whole has a particular relevance to our project, because the model is restricted to using functional forms that are very close to those specified by economic theory. Still more relevant for our purposes is the fact that the condensed model is very close in size to the model generated in this study. For these two reasons the Growth Project model has been particularly useful in assigning parameter values to the model developed here.

Overall the Growth Project model is a very impressive one, both in terms of its theoretical basis and in terms of its econometric execution. There are, of course, general theoretical criticisms which can be made at the econometric level. Most equations are estimated independently and the estimation technique varies between equations. Also, the data base is not consistent for the whole model. However, given the size of the model these must be seen as theoretical criticisms only, and it is very hard to envisage any practical alternative.

Within the framework of the Cambridge Multisectoral Domestic model there are a number of smaller sub-models, covering certain aspects of the economy in greater detail. One of these is the energy sub-model, which basically shows the structure of energy demand in the UK, and forecasts the demand for seven main fuels. The supply side of the fuels is not, however, analysed within this model, and so no depletion policy is developed for North Sea oil.

Finally, the Growth Project model is set apart from the other models by its adherence to economic theory and, as has already been stressed, by its aim of modelling the medium term rather than the extreme short term. This approach generates a model which is clearly different in terms of its short-term properties, but is at the same time much more reliable over the long term and much more robust with respect to structural changes in the economy. This feature makes the model a more powerful tool for policy simulations, both because the longer-term projections of the model will be more meaningful and because it will respond more realistically to the simulated policy changes.

(iv) *The London Business School Model*

The London Business School model is one of the smaller of the large models, with well under 200 endogenous variables. The model has a long history, dating back to 1966, and an earlier version of the present model was adapted by the Bank of England and used to found their own model. This model will be reviewed under a separate heading.

Originally, the London Business School intended that forecasting should be carried out by a simple, small model while a more complex model was developed for simulation purposes. In fact, this aim was not achieved. As the forecasting model developed, it became larger and the problem of updating the simulation model proved too great to be practical. But as the forecasting model grew in size so the need for a separate simulation model disappeared. By the early 1970s the forecasting model had in effect become the one central model of the Business School, although it could be extended for simulation by the use of various small sub-models.

During the middle and late 1970s the model underwent fairly drastic changes, mainly as a result of its poor performance during the oil crisis earlier in the decade. This process has caused it to diverge markedly from its offshoot model at the Bank of England, as well as from most of the other large models of the British economy.

At the most aggregate level the model is a fairly conventional income-expenditure one, like all the other large models. Quantities, in particular output, are predominantly demand-determined. Consumption is a function of lagged consumption, disposable income, wealth and various dummy variables. Investment is a function of exogenous public investment, the rate of change of output, lagged output and a time trend. Exports are a function of time and lagged exports. Imports are a complex function of the money stock, lagged imports, the change in output, various dummies and a number of productivity terms.

The North Sea sector is specified in a moderately detailed way although the direction taken by the model is purely one of an exogenous North Sea oil sector affecting the rest of the economy. This is done through additional taxation and balance of payments effects. The model does not concern itself with depletion policies, an aspect that is broadly in line with the other models.

The London Business School model diverges from the others, however, in the areas of prices, money and exchange rates. Generally, most of the other models have adopted what can be classified as a Keynesian view of the role of money-price setting and the exchange rate: money has only limited interest rate effects, prices are set mainly on the basis of cost mark-up procedures and the exchange rate cannot easily be handled. The London Business School model on the other hand adopts a fairly sophisticated international monetarist view of the determination of these variables. In particular the model is based on a theory of inflation (Ball and Burns)[3] which sees the long-term inflation and exchange rates as determined by the relative changes in the money supplies of different countries at an international level. Finally, as far as estimation of the model is concerned, wide use is made of explanatory variables in the equations.

It is the monetarist approach of the London Business School model that distinguishes it from other macroeconomic models, on the theoretical side. On the practical side, the model is basically suitable for short-term forecasts. Its longer-term properties will depend essentially on the accuracy of its theoretical basis as opposed to the more conventional Keynesian type. Overall, the London Business School model has very little relevance for the one developed in this study. Their primary objectives are different from ours as is the balance between theory and empirical investigation.

(v) *The Bank of England Model*

The macroeconomic model of the Bank of England originated in 1972–1973 when the Bank acquired a version of the London Business School model. Since that time this model has been revised and expanded to such an extent that it must now be treated as an entirely independent one. The purposes for which the model is designed are primarily short-term and medium-term forecasts and simulations. As might be expected of a central bank model, it is meant to have a substantial, highly detailed financial sector although the development of this side of the model is not yet as far advanced as the Bank would wish. (This is mainly due to the fact that the original London Business School model almost completely lacked a financial sector.) There is, however, already a fairly complex flow-of-funds matrix, specifying some thirty variables inside the model.

With some 500 endogenous variables, the Bank model is probably one of the larger ones among the current major British economic models, but it is hard to compare the sizes of models directly. The main reason why the Bank model is so large is because it contains many separate government instruments that are all specified independently. For example, there are over thirty separate types of taxes specified within the model, all of which can be manipulated independently in policy simulations. This form of specification has led inevitably to a large degree of disaggregation wherever these instruments affect the economy. This is at the cost of making the model more cumbersome.

Each individual equation of the model is estimated by means of the ordinary least squares estimation method. Where it is felt that this procedure gives unreasonable results, the estimates of the coefficients have been imposed judgmentally, which on the whole enhances the flexibility of the model.

The Bank of England model is basically a Keynesian one, in the sense that quantities are generally demand-determined. The level of output is determined through an aggregate demand function. Consumption is a function of income and net liquid assets, forming a wealth effect. Investment is determined through a fairly straightforward set of accelerator equations. This is where their model differs from ours.

The North Sea oil-and-gas sector is included within the body of the model in a number of separate equations, as an argument determining the level of exports, for example. However, the model largely takes the North Sea sector to be exogenous (both in terms of price and quantity) and so the modelling effort is devoted only to showing how the North Sea might affect the economy. This differs markedly from our model, since we endogenize the oil sector and determine the optimal depletion policies within the model.

The Bank of England Model is an extremely powerful tool of short-term forecasting and general policy simulation. But, from the point of view of this project, it has very little to offer. For example, when a simple consumption function is being considered and assigned values, an equation such as the one used by the Bank, containing thirteen variables and one constant term, has little direct relevance for a long-term problem such as that formulated in this study.

(vi) *The Cambridge Economic Policy Group Model*

The Cambridge Economic Policy Group model is one of the smallest of the large econometric models, with only 136 endogenous variables. It is in many ways quite distinct from most of the other models both in construction and methodology. The basic objective of this model is to produce forecasts over a medium-term time horizon, normally about ten years. The model operates on an annual basis, unlike most of the other models which are quarterly.

As far as model construction is concerned, the Cambridge Economic Policy group have tried to make as few a priori restrictions as possible. Instead they have allowed the estimation procedure to obtain the best possible fit. The general functional form around which all their relationships are formulated is:

$$g(X_{it}) = f(X_{jt,} Z_{kt}) + \alpha_i + B_i T + U_t$$

where $g(X_{it})$ is either X_{it} or ln X_{it}

The X_{jt} are the other dependent variables and Z_{kt} are the exogenous and lagged endogenous variables. $B_i T$ is a time trend.

The Policy Group model is basically a Keynesian one, and quantities are, broadly speaking, demand-determined. Aggregate output is determined by an aggregate demand function where consumption is a function of a trend value and current and lagged disposable income; investment is a function of a trend and disposable income, and exports are a function of a trend, world-trade levels and relative prices.

The North Sea sector enters into the model as a determinant of other sectors, and both the price of oil and the quantity produced are taken as exogenous. The value of output is estimated as, simply, output multiplied by price, minus some exogenous costs. Government revenue from the North Sea is given simply by the value of North Sea oil output multiplied by an exogenous percentage tax rate.

From the point of view of our study the Cambridge Policy Group model too, has very little direct relevance. The model developed in our research is based very firmly on theoretical economic reasoning; the Policy Group model on the other hand has been developed from the opposite viewpoint. Furthermore, the model does not concern itself with depletion policies. Thus the areas of close compatabilities are very limited.

3.3 Conclusions

In principle a model which is designed mainly for forecasting short-term macroeconomic trends, such as the Treasury model, will be structurally very different from a model designed to forecast longer-term trends. Also, the results generated from any model can be very sensitive to minor respecifications of the model itself.[4] This point can be illustrated by running simulations on different versions of any of the models outlined in this chapter.

Our survey enables us to conclude that the current state of model-building is moderately satisfactory in terms of short-term models, but that there are no satisfactory long-run models for the UK able to determine optimal policies for investment and oil extraction simutaneously. Even the very aggregate and long-run version of the Cambridge Growth Project model does not calculate any depletion profiles for the North Sea oil. (Indeed it was not designed to do so, and this is not a criticism of the model.)

In the absence of an existing model that would fulfill the twofold objective of investment and depletion simultaneously, it became evident that a specific model had to be developed for this purpose. Thus the macroeconomic model specified in chaper 2 was developed to cater for the long-term nature of the problem, and the optimal control technique was applied to determine the optimal investment and depletion policies.

4 Taxation and Estimation of UK Oil Revenue

4.1 Introduction

This chapter contains a detailed description of the UK oil taxation system (as it stands at the end of April 1982) and its historical evolution. Also, some empirical estimation is made of the likely magnitude of government revenues accrued from the North Sea oil taxation. The material is presented in the following way. First, the oil taxation system in force until the end of 1982 is described, together with the new arrangements that take effect from 1 January 1983. Second, an account is given of the changes that the system has undergone since its inception. Third, the magnitude of the revenue is measured for the 1976–2000 (which coincides with our planning period in this book) accruing from the twenty-four fields whose proven recoverable reserves serve as a basis for the initial oil stock that is assumed in our optimization calculations appearing in chapter 5.

We have been especially careful to prepare the material for this chapter in such a way that it is as up-to-date as possible – if only because the oil taxation in Britain seems to be subject to continual revisions! Moreover, some of the significant variables such as the price of oil, the dollar/sterling exchange rate, the rate of oil extraction and the quantity of proven reserves can change as a result of different market and macroeconomic conditions. The latest taxation changes that occurred before this book went to press were announced in the March 1982 Government Budget by the Chancellor of the Exchequer. This necessitated some revision of the original chapter to incorporate these changes. In the process, we also used the latest figures for all the other variables entering our computation of government revenue. These included the estimate of the proven oil reserves, which is, in fact, slightly less than the original figure used in chapter 5 as our stock of oil reserves.

4.2 The Structure of the UK Oil Taxation system (Until The End of 1982)

The UK oil taxation system has evolved considerably since its inception in the mid-1970s. By 1982, four different types of taxes could be identified, namely, royalties, supplementary petroleum duty, petroleum revenue tax and corporation tax, to which the oil companies were liable. These sources of revenue, together with their main features, arc each described in turn.

Royalties

The provision for royalties to be paid at a rate of 12½% on the 'value' of oil production from the North Sea was established by the Continental Shelf Act, 1964.[1] Based on discoveries made under the first four licensing rounds (1964, 1965, 1970 and 1971/2) this value was taken to be the well-head value. Deductions could be made from the landed or sales value to allow for some initial treatment and transportation costs. The effective royalty rate was thus lower as a percentage of sales value, with the precise figure varying from field to field. On production from fields licensed under the fifth (1976/77), sixth (1978/79) and seventh (1980/81) rounds the appropriate 'value' was changed to the landed value.

No changes were made in the 1982 Budget. Liability for royalties is calculated on the basis of the total value of production for six-month chargeable periods ending 30 June and 31 December of each year. Payment is due within two months of the end of the period.

Supplementary Petroleum Duty

Supplementary Petroleum Duty (SPD) was initially proposed by Sir Geoffrey Howe, the Chancellor of the Exchequer, in November 1980 and was formally introduced five months later as part of the 1981 Budget.[2] This tax was levied at a rate of 20% on gross production revenues subject to an annual exempt allowance of one million tonnes of oil for each field. No part of this annual allowance remaining unused in a particular year could be carried forward. SPD is calculated twice a year for the same chargeable periods used in the derivation of royalty liabilities—the first payment was due on the 1 September 1981 for production in the first half of that year. Thereafter payments were scheduled in monthly instalments to bring them more into line with actual production.

Starting at the beginning of the fourth month of any particular chargeable period (that is 1 October or 1 April), five separate monthly payments were due, each equal to one-fifth of the liability for the previous period. A sixth payment (or rebate), due two months after the end of the chargeable period (on 1 March or 1 September), consisted of an 'adjustment' after an accurate assessment of the liability for that period had been made. No interest was charged (or paid) should the amount 'pre-paid' have failed to cover (or exceed) the accurately assessed liability.

In the first instance, these arrangements for the payment of supplementary petroleum duty only applied to the eighteen-month period from the beginning of 1981 to the middle of 1982. The tax proved to be unpopular with the oil companies and, in October 1981, two separate groupings—as well as several outside observers of the industry—submitted proposals to the government for the reform of the whole taxation system. These two alliances—the 29-member UK Offshore Operators Association, which included the largest North Sea operators, and the Association of British Independent Oil Exploration Companies, which represented 38 smaller companies, both called for the abolition of SPD, because, they maintained, it severely affected the profitability of marginal fields.

The Chancellor subsequently went some way to meeting these demands in his Budget of 1982.[3] Liability to SPD was to be extended for a further six months, until the end of 1982, but thereafter the tax was to be abolished. However, other changes to the taxation system proposed at the same time seemed to resurrect the main elements of SPD in an alternative guise of advance payments of petroleum revenue tax. (See the section 4.3 for a description of the tax system from 1983 onwards). Table 4.1 gives details of payments of the supplementary petroleum duty to which the oil companies were liable.

Petroleum Revenue Tax

Petroleum Revenue Tax (PRT) was introduced by the Oil Taxation Act (1975)[4] and is charged on profits made by oil companies from their activities in the North Sea. Like the more recent SPD, PRT is calculated on a six-monthly basis for each field separately so that development costs of one field may not be used to defer payment of tax on the profits of another. PRT has always been assessed at a standard rate on the basis of total revenues, less royalty and supplementary petroleum duty payments, operating costs and certain allowances. However, the rate, the allowances and the timing of the payments have all been changed substantially over the past few years. These changes are detailed in section 4.4. The terms and conditions given here relate to the tax regime as was confirmed in the March 1982 Budget. The main allowances are as follows:

(i) Total capital expenditure on the field 'uplifted' by 35%. The 'uplift' is to compensate companies for the fact that interest payments are not allowable as a deduction when calculating the PRT liability. Capital expenditure incurred after a field has reached payback, that is, after the chargeable period when cumulative

Table 4.1 Payments of supplementary petroleum duty

1 January - 30 June 1981	first chargeable period
1 July - 31 December 1981	second chargeable period
1 January - 30 June 1982	third chargeable period
1 July - 31 December 1982	fourth chargeable period
1 September 1981	payment for first chargeable period
1 October 1981 1 November 1981 1 December 1981 1 January 1982 1 February 1982	payment in each month of one-fifth of the amount paid in respect of the first chargeable period
1 March 1982	payment for second chargeable period less amount already paid in advance
1 April 1982 1 May 1982 1 June 1982 1 July 1982	payment in each month of one-fifth of the amount paid in respect of the second chargeable period.
1 September 1982	payment for third chargeable period less amount already paid in advance
1 October 1982 1 November 1982 1 December 1982 1 January 1983 1 February 1983	payment in each month of one-fifth of the amount paid in respect of the third chargeable period
1 March 1983	payment for fourth chargeable period less amount already paid in advance

incomings first exceed cumulative outgoings, will not qualify for 'uplift' but will, nevertheless, still be deductable at historical cost.

(ii) An annual oil allowance of 0.5 million metric tonnes per field (subject to a cumulative total of 5 million tonnes over the life of the field).

The PRT liability was then 70% of the base, so calculated subject to two further conditions:

(iii) A 'safeguard' provision which guarantees that the PRT payable in any year will not reduce the return on a field to less than 30% of capital expenditure measured on the basis of historical costs. Only capital expenditure which qualifies for 'uplift'—that incurred before payback is reached—is to be used in the 'safeguard' calculation. The time during which the 'safeguard' provision may be applied is limited to the period when the field has reached payback plus half of the time from production start-up to payback.

(iv) A 'tapering' provision to ensure that the PRT charge will not be greater than 80% of the amount (if any) by which the return exceeds the 'safeguard' level. In other words, no PRT is payable if the adjusted profit—gross revenues less royalties, SPD and operating costs—is less than 30% of accumulated capital expenditure. The 'tapering' provision applies if two conditions are fulfilled: (a) if the adjusted profit is greater than 30% of capital expenditure, and (b) if the field would be liable (under the basic 70% rate) to more than 80% of the amount by which the return exceeds the safeguard level.

Normal payments are due on account written two months after the completion of each chargeable period ending on the last days of June and December each year. However, some advance payment is required six months beforehand, the amount of which is the greater of:

15% of the assessed liability for the previous-but-one chargeable period;
15% of the payment on account for the previous chargeable period.

The advance will then be deducted from the normal payment due on account six months later.

Corporation Tax
This tax is payable annually at the standard rate of 52% on net

income after deductions for royalties, SPD, PRT, interest payments and the accelerated depreciation of capital expenditure. In contrast to royalties and SPD, which are levied on revenues, and PRT which is a profits-based tax paid on a field-by-field basis, corporation tax is calculated on a company basis. Hence, capital costs in fields under development can be used to offset profits from fields already in production. A 'ring fence' has, however, been erected for tax purposes around operations on the Continental Shelf. Capital allowances and losses applicable to company activities outside the 'ring fence' may not be written off against profits from the North Sea. Payment of corporation tax on new trades is due nine months after the end of the accounting period in which the liability was incurred. Actual payment may be made up to thirty days after the due date without incurring any interest liability.

4.3 The Structure of The UK Oil Taxation System (from 1983 onwards)

In the Budget of 1982, the Chancellor of the Exchequer proposed various changes[5] to the oil taxation system described in the last section. The main alteration was that supplementary petroleum duty was to be abolished with effect from 31 December 1982. In its place, the rate of petroleum revenue tax was increased from 70% to 75% and a new system of advance payments was introduced, both with effect from 1 January 1983. These advance payments of petroleum revenue tax (APRT) will be levied at a rate of 20% on gross revenues after deducting a tax-free allowance of one million tonnes of oil per year. In these respects, the new advance payments of PRT are indeed similar to their predecessor, supplementary petroleum duty. However, whereas the latter was a separate tax which was an allowable deduction against petroleum revenue tax liability only in the period in which it was incurred, the former is different in that it is an acceleration of existing PRT payments and will be allowable against both current and future PRT liabilities.

The arrangements for the actual payment of PRT (including APRT) were also changed in order to secure a smoother cash flow to the Exchequer. With effect from the second chargeable period of 1983, six equal monthly payments will be required, the first due two months after the start of the period—initially on 1 September 1983, and subsequently on 1 March and 1 September of each year, which together will constitute 75% of the previous period's liability. This liability is defined as being inclusive of any APRT payments. The balance due on account will be paid two months after the end of the chargeable period in question, by which time an accurate assessment of the liability will have been made. Hence, in any typical year say 1984, payments of PRT (including APRT) as set out in Table 4.2 can be calculated. Conditions relating to the payment of royalties and corporation tax, on the other hand, were not affected by the March 1982 Government Budget.

4.4 Changes in the Legislation Relating to the North Sea Oil Taxation System

The previous two sections contain descriptions of the oil taxation system as it stood during 1982 and the changes introduced to take effect from 1983 onwards. Even before then, however, governments—both Labour and Conservative—had made a number of important alterations to the system, particularly in the format of petroleum revenue tax, since its first formulation.

When production first started in the North Sea, government tax revenue was collected through only three sources—royalties, petroleum revenue tax and corporation tax. Supplementary petroleum duty had yet to be conceived. PRT liabilities were calculated every six months and payment was due four months later. The 'uplift' on capital expenditure was set at 75% and there were no time limitations on which expenditure qualified either for 'uplift' or for the 'safeguard' provision which was calculated on an annual basis. The annual oil allowance was one million long tons (subject to a cumulative total of ten million tons) and the PRT rate was fixed at 45%. No advance payment was required. This initial tax system reflected the cautious attitude being adopted by the authorities towards North Sea development at the time.

By 1978, however, it had become clear that the oil companies were obtaining very large profits from their North Sea activities. Accordingly, on 2 August 1978,[6] the Labour government announced a number of proposed changes in petroleum revenue tax which were designed to increase the state's share of revenues:

(i) The 'uplift' on capital expenditure was reduced from 75% to 35% for qualifying expenditure under contracts made after 2 August 1978

(ii) With effect from 1 January 1979 the annual oil allowance was reduced from one million long tonnes to 0.5 million metric tonnes subject to a new cumulative limit of five million tonnes per field.

Table 4.2 Payments of petroleum revenue tax during a typical year with 1983 and 1984 as examples

1 September 1983 1 October 1983 1 November 1983 1 December 1983 1 January 1984 1 February 1984	'Smoothed' payment of PRT for the chargeable period July – December 1983—six payments each equal to one sixth of 75% of the assessed PRT liability for the chargeable period January – June 1983
1 March 1984	Payment of the balance due on account for the chargeable period July – December 1983, equal to the assessed PRT liability for the chargeable period July – December 1983 *less* 75% of the PRT liability for the chargeable period January – June 1983 (the six 'smoothed' payments detailed above); *plus* the first 'smoothed' payment of PRT for the chargeable period January – June 1984, equal to one sixth of 75% of the assessed PRT liability for the chargeable period July – December 1983
1 April 1984 1 May 1984 1 June 1984 1 July 1984 1 August 1984	the remaining five 'smoothed' payments of PRT for the chargeable period January – June 1984
1 September 1984	Payment of the balance due on account for the chargeable period January – June 1984, equal to the assessed PRT liability for the chargeable period January – June 1984 *less* 75% of the PRT liability for the chargeable period July – December 1983 (the six 'smoothed' payments detailed above); *plus* the first 'smoothed' payment of PRT for the chargeable period July – December 1984, equal to one sixth of 75% of the assessed PRT liability for the chargeable period January – June 1984
1 October 1984 1 November 1984 1 December 1984 1 January 1985 1 February 1985	The remaining five 'smoothed' payments of PRT for the chargeable period, July – December 1984

(iii) The rate of PRT was increased from 45% to 60%—again with effect from 1 January 1979.

These proposals became law as part of the Finance Act.[7]

The second amendment to the North Sea taxation system was announced by the new Conservative government on 15 November 1979.[8] A two-month advance in the date for collection of petroleum revenue tax was introduced, thus bringing the arrangements into line with those for collecting royalties. PRT payments which had been due on 1 May and 1 November for the previous chargeable period were now due by 1 March and 1 September respectively. The first payments under this new regime were to be collected on 1 March 1980.

The third series of alterations[9] to the offshore tax regime were those announced in the course of the budget of 26 March 1980, which then became law as part of the 1980 Finance Act. The increase in the basic rate of PRT from 60% to 70% for chargeable periods ending after 31 December 1979 had been widely expected by the industry in the light of fast-rising oilprices. However, the decision to introduce advance payments of a proportion of the tax by six months was something of a surprise.

And one year later, in the Budget of 1981, yet more changes were proposed[10] which completed the evolution of the system to the form described in section 4.2. The introduction of supplementary petroleum duty had been introduced in November 1980, but the extent of the measures designed to tighten up reliefs against PRT were somewhat unexpected. It appeared that the government had been anxious to discourage additional capital expenditure after a field had become well-established as a producer. Hence, they introduced the restrictions on 'uplift' and 'safeguard' and changed the calculation of 'safeguard' from an annual basis to a six-monthly one in order to provide a more even pattern of payments. Finally, came the revisions made in petroleum revenue tax, effective from the 1 January 1983. All in all, the five major overhauls of the tax system within four years is a record which reflects the uneasy atmosphere that has been created around the oil business in the North Sea.

4.5 Estimation of Government Revenues

Forecasting the magnitude of future government income from North Sea oil is an undertaking which can be best described, with considerable understatement, as being fraught with difficulties. There are so many variables to take into account and changes in any of them may dramatically affect both the level and the timing of the tax revenue flows. More specifically, even if all potential sources of oil from the North Sea could be identified, and the reserves recoverable therefrom ascertained with certainty, accurate prediction would still require a detailed knowledge of the following:

1 *Government policy towards the depletion of North Sea oil reserves.* The government can influence the rate of depletion through its powers to grant licenses for exploration and then approval for the development of established reservoirs.

2 *Capital and operating expenditure costs.* For each field for which development approval has been granted, time profiles are needed for production rates and the associated capital and operating expenditure costs.

3 *The ownership of each field.* Corporation Tax, unlike the other three oil taxes (namely, royalties, supplementary petroleum duty and petroleum revenue tax), is payable on a corporate basis. Actual payments of corporation tax on a particular field thus depend on the activities of the participating companies elsewhere in the North Sea.

4 *Equity financing and interest payment.* The level of equity attributable to each field and the interest paid on any loans raised.

5 *Price of oil.* The dollar price of oil for each of the various grades of oil produced.

6 *Inflation rate.* The likely future rate of inflation of both capital and operating costs.

7 *Exchange rate.* The dollar/sterling exchange rate.

8 *Taxes on North Sea.* Details of the North Sea oil taxation system and its likely future changes.

Much of this information cannot be forecast with any degree of certainty. The full extent of North Sea oil reserves is not known. Both production from reservoirs already discovered, and further exploration and exploitation of other sources, depend on many factors not the least of which will be the future course of the price of oil. As is well known, the price of oil is notorious for its uneven course over time (see Figure 5.1 in the next chapter) and it would be folly to place too much reliance on any one particular forecast. Similarly, inflation and exchange-rate movements are also difficult, if not impossible, to predict accurately. The pattern of ownership and the financial arrangements will vary with specific cases. Production rates, capital and operating costs are subject to the vagaries of many an unpredictable influence, including the weather. Last, but not least, there is the overriding question of government policy towards the tax regime and the size of the take. This has been subject to many changes in the past, and it will very likely see more in the future.

Our list of the uncertainties surrounding the North Sea oil industry is by no means exhaustive, but it illustrates the difficulty of trying to provide reasonably sensible estimates of government income. There is a wide range of possible scenarios for each of the many variables within the system (high inflation rate, low inflation rate, and so on), and this in-built lack of precision should be both taken into account in the analysis itself and borne in mind when the numerical results are interpreted.

Our Analysis

To simplify the problem, we have limited our analysis to those twenty-four fields for which development approval had already been granted by the end of 1981. These fields are, in alphabetical order, Argyll, Auk, Beatrice, Beryl, Brae, Brent, Buchan, Claymore, Cormorant,[11] Dunlin, Forties, Fulmar, Heather, Hutton, Hutton North West, Magnus, Maureen, Montrose, Murchison UK, Ninian, Piper, Statfjord UK, Tartan and Thistle. Between them these fields were expected to yield 11,949 million barrels[12] of oil (see Table 5.14 in the next chapter). All of them were licensed under the first four rounds of licensing and, hence, their effective royalty rates are less than 12½% of sales value. This limitation means that the revenue estimates given for these fields should only be regarded as forecasts of actual government income for the years up to, say, 1987. Thereafter, other fields, for which development approval has not yet been granted, may also be in production and helping to fill the government coffers. The stockbrokers Wood, MacKenzie & Co have drawn up a list of fields that might easily have received official sanction by the end of 1983 (see Table 4.3).

The 'base' case

Computer simulations were undertaken for the chosen twenty-four fields, for the years 1976—2000, under a number of different scenarios in order to examine how sensitive the results were to alternative assumptions on oil price, inflation rate, exchange rate, depletion rate, etc. But first a 'base' case was considered where the following assumptions were made:

(i) Each field was treated as if it were privately owned. This, of course, neglects the role of the British National Oil Corporation as a nationalized industry, but is a reasonable simplification if one considers that the state-owned company was given powers under the Petroleum and Submarine Pipelines Act (1975) to operate as a commercial concern.[13] BNOC has been thus far, at least in theory, an independent entity.

(ii) Corporation-tax liabilities were calculated on a field-by-field rather than on a corporate basis because of the lack of information relating to the capital expenditure plans of individual companies. Such a procedure will, however, predict an earlier payment of corporation tax than is likely to be the case in practice. This should be taken into account when analysing the estimates of the absolute levels of government revenues. However, relative levels—on different assumptions of the price of oil, for instance—should be little affected and, as such changes are the main focus of analysis in this chapter, the adopted procedure should lead to little distortion of the results.

Table 4.3 North Sea oil fields which may receive development approval by the end of 1983

Field	*Estimated Recoverable Reserves (barrels/million)*	*Estimated Production Start*
Alwyn North	130	1987
Andrew	100	1986/87
North Brae	150 – 200	1987
Bruce	150 – 180	1988
Caber	100 – 150	1987/88
Clyde	125	1987
T Block	200 – 350	1987/88
Term	120 – 150	1988
14/20	30	1983/84

Note: This table does not include all the new developments which might be undertaken if the following extensions to existing reservoirs take place: West Argyll, Claymore Area, Central Cormorant, East Forties, North West Heather and South Ninian.

Source: Wood, Mackenzie & Co., *North Sea Report*, No.100 (28 August 1981) pp. 3-5.

(iii) A general level of gearing of 25%[14] has been assumed, and interest on loans has been charged at a rate equal to the assumed rate of inflation, i.e. a real rate of interest of zero per cent.

(iv) The general rate of inflation is based on the UK wholesale price index for the period 1972–81. An annual rate of 10% has been postulated for 1982. Thereafter a rate of 8% has been assumed.

(v) The price of oil from all fields has been assumed to be the same as that for Forties marker crude. This assumption slightly overestimates revenues from fields with a lower quality crude (e.g., Claymore, Heather, Ninian, Piper and Tartan) but also underestimates revenues from fields with a higher quality crude (e.g., Beryl and Montrose). The underestimated values are likely to partially offset the overestimated values so that the overall distortion due to this pricing system is likely to be negligable. Actual oil prices are used for the years 1975 up to March 1982. Thereafter the price is taken to be constant until the end of 1982 and the real sterling price of oil to rise subsequently at 6% per annum.

(vi) The dollar/sterling exchange rate is based on actual values for the period 1972–81. It is forecast to have an average level of $1.85 = £1 for the first six months of 1982 and $1.90 = £1 for the last six months. It is then assumed to remain at this rate.

(vii) Oil production from the British sector of the North Sea is forcast at about 2.1 million barrels per day in 1982, rising to a peak of 2.5 million barrels per day in 1985, and falling slowly thereafter up to the turn of the century.

(viii) Percentage government take is calculated as the nominal sum of royalties, supplementary petroleum duty, petroleum revenue tax and corporation tax payments as a proportion of net revenues (total revenues less capital and operating costs) over the entire life of the fields considered.

(ix) Calculations were performed in nominal sterling values for six-month chargeable periods ending on the last days of June and December each year. Figures relate to payments rather than accrued liabilities and are attributed to calendar years (in preference to the normal practice of fiscal years) to maintain consistency with the analysis in the rest of the book.

The calculations were carried out on a field-by-field basis but only aggregate estimates of government receipts are given here. Tables 4.4 and 4.5 give 'base' case projections in current and constant 1975 prices respectively. As can be seen from Table 4.4, actual tax revenues should peak towards the end of the 1980s and fall off very slowly afterwards. By far the largest component of the total take is the income provided by petroleum revenue tax. This accounts for some 59% over the whole period—a proportion which rises to over 70% of the annual take during the later years of the century. This is not surprising, since Petroleum Revenue Tax (PRT) is essentially a tax on profits and, after the mid-1980s, most of the fields under consideration will be well established and thus generating substantial net income flows. In contrast, payments of royalties and supplementary petroleum duty (until its abolition), which are both taxes levied on revenues, are more correlated to the production profile. There, the government take is 82%.

Table 4.4 Total government tax take from North Sea oil, 1976—2000 ('base' case—current prices)

Year	Royalties (£ million)	Supplementary Petroleum Duty (£ million)	Petroleum Revenue Tax (£ million)	Corporation Tax (£ million)	Total Government Take (£ million)
1976	24	0	0	0	24
1977	165	0	0	9	174
1978	278	0	182	316	775
1979	419	0	701	629	1,749
1980	825	0	1,938	831	3,594
1981	1,100	1,497	2,407	997	6,001
1982	1,392	2,244	1,493	1,116	6,244
1983	1,447	365	3,205	1,744	6,761
1984	1,775	0	4,386	2,756	8,916
1985	2,112	0	5,755	3,343	11,210
1986	2,360	0	7,173	3,747	13,279
1987	2,453	0	7,827	4,109	14,388
1988	2,426	0	8,994	4,634	16,054
1989	2,358	0	8,141	3,908	14,406
1990	2,312	0	7,947	4,117	14,376
1991	2,268	0	8,502	3,989	14,758
1992	2,210	0	8,379	3,391	13,981
1993	2,188	0	9,005	3,208	14,402
1994	2,182	0	9,039	2,794	14,015
1995	2,149	0	9,212	2,616	13,977
1996	2,135	0	9,352	2,256	13,742
1997	2,146	0	9,589	2,102	13,837
1998	2,037	0	9,143	1,927	13,107
1999	1,893	0	8,828	1,563	12,283
2000	1,844	0	8,567	1,541	11,951

**Tax take applies to proven reserves (as it stood in 1982) – see text for details.*

Such nominal figures, although corresponding to forecasts of actual payments to the Exchequer, nevertheless conceal the real flow of resources. Table 4.5 presents the same information as Table 4.4, but in constant 1975 prices. This shows that, in real terms, tax revenues should actually peak in 1988 and then decline quite quickly as more and more fields are depleted. The effect is demonstrated strikingly in Figures 4.1 and 4.2. The total tax take over the twenty-five year period is £56,965 million in 1975 prices, which corresponds to about 57% of gross revenues.

Table 4.5 Total government tax take from North Sea oil, 1976—2000 ('base' case—constant 1975 prices)

Year	*Royalties* (£ million)	*Supplementary Petroleum Duty* (£ million)	*Petroleum Revenue Tax* (£ million)	*Corporation Tax* (£ million)	*Total Government Take* (£ million)
1976	20	0	0	0	20
1977	116	0	0	6	122
1978	182	0	116	202	500
1979	242	0	402	351	995
1980	411	0	964	406	1,781
1981	496	661	1,089	440	2,685
1982	571	921	614	448	2,554
1983	547	141	1,210	648	2,546
1984	622	0	1,539	948	3,108
1985	685	0	1,864	1,064	3,614
1986	709	0	2,158	1,105	3,972
1987	683	0	2,181	1,122	3,986
1988	626	0	2,325	1,171	4,122
1989	563	0	1,949	915	3,427
1990	511	0	1,760	892	3,163
1991	464	0	1,744	800	3,009
1992	419	0	1,592	630	2,641
1993	384	0	1,582	552	2,518
1994	355	0	1,472	445	2,271
1995	324	0	1,389	386	2,098
1996	297	0	1,304	308	1,910
1997	277	0	1,238	266	1,781
1998	244	0	1,095	226	1,564
1999	209	0	978	169	1,357
2000	189	0	879	155	1,222

**Tax take applies to proven reserves (as it stood in 1982) – see text for details.*

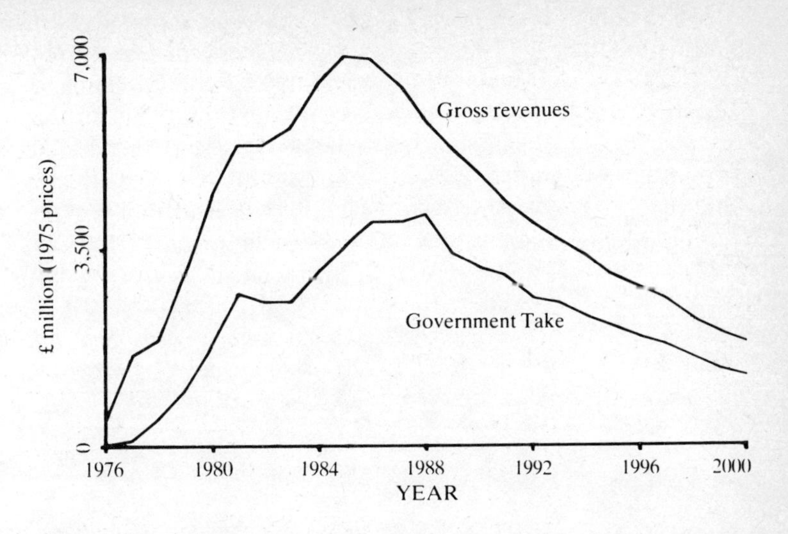

Figure 4.1 Graph of gross revenues and government tax take from North Sea oil, 1976—2000 (constant 1975 prices corresponding to Table 4.5)

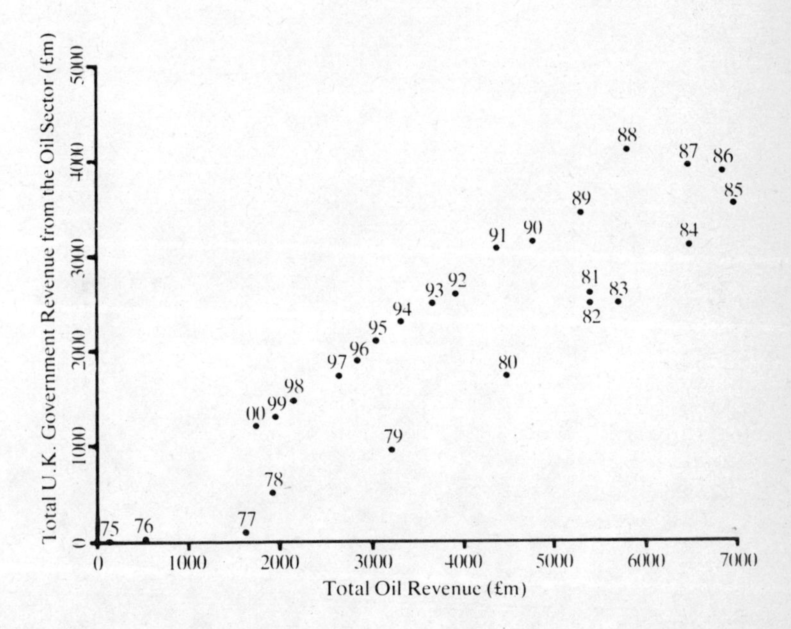

Figure 4.2 Government taxation revenue from the UK oil sector in constant 1975 prices

Changing the parameters
Such are the results of the 'base' case. Five sets of simulations were then carried out to examine the sensitivity of these estimates to changes in the following parameters: the real price of oil, the rate of UK inflation, the dollar/sterling exchange rate, the rate of oil depletion, and capital and operating expenditure.[15]

The effect of varying the real price of oil is unsurprising. As can be seen from Table 4.6, the higher the rate of increase assumed, the higher are both the annual flows and the overall tax take. If the real sterling price of oil is taken to remain constant from the end of

Table 4.6 The effect on total government tax take of different assumptions regarding the real sterling price path of oil
£ million — constant 1975 prices

Year	*Total government tax take when, from the forecast level at the end of 1982, the average real sterling price of oil is assumed to rise at:*			
	0% p.a.	*3% p.a.*	*6% p.a. ('base' case)*	*8% p.a.*
1976	20	20	20	20
1977	122	122	122	122
1978	500	500	500	500
1979	995	995	995	995
1980	1,781	1,781	1,781	1,781
1981	2,685	2,685	2,685	2,685
1982	2,554	2,554	2,554	2,554
1983	2,504	2,525	2,546	2,560
1984	2,882	2,974	3,108	3,198
1985	2,999	3,302	3,614	3,866
1986	3,017	3,504	3,972	4,250
1987	2,789	3,341	3,986	4,353
1988	2,824	3,386	4,122	4,765
1989	2,314	2,752	3,427	3,934
1990	1,720	2,367	3,163	3,962
1991	1,550	2,099	3,009	3,601
1992	1,224	1,853	2,641	3,355
1993	1,059	1,718	2,518	3,138
1994	888	1,498	2,271	2,950
1995	763	1,306	2,098	2,792
1996	582	1,111	1,910	2,608
1997	478	994	1,781	2,515
1998	391	834	1,564	2,252
1999	307	693	1,357	1,997
2000	245	596	1,222	1,840
TOTAL	37,192	45,510	56,965	66,593

1982 onwards, the overall tax take amounts to £37,192 million—a drop of 35% compared with the 'base' case of a 6% increase. A 3% rate of increase yields a total take of £45,510 million (20% down on the 'base' case) whereas an 8% rate of increase leads to 17% more in government revenues.

The objective in simulating for different rates of UK inflation was to examine the extent to which capital allowances might be devalued and tax revenues thereby affected. Hence, it was decided that thc nominal price of oil should be increased concomitantly with the postulated rate of inflation so that the real sterling rate of

Table 4.7 The effect on total government tax take of different assumptions regarding the rate of UK inflation

£ million — constant 1975 prices

Year	*Total government tax take when, from the end of 1982, the rate of UK inflation is assumed to be:*		
	4% p.a.	*8% p.a. ('base' case)*	*12% p.a.*
1976	20	20	20
1977	122	122	122
1978	500	500	500
1979	995	995	995
1980	1,781	1,781	1,781
1981	2,685	2,685	2,685
1982	2,554	2,554	2,554
1983	2,599	2,546	2,496
1984	3,181	3,108	3,041
1985	3,587	3,614	3,641
1986	4,035	3,972	3,901
1987	3,980	3,986	3,926
1988	4,243	4,122	4,104
1989	3,492	3,427	3,361
1990	3,183	3,163	3,273
1991	2,874	3,009	3,002
1992	2,697	2,641	2,642
1993	2,546	2,518	2,452
1994	2,338	2,271	2,241
1995	2,115	2,098	2,074
1996	1,906	1,910	1,890
1997	1,791	1,781	1,764
1998	1,594	1,564	1,552
1999	1,366	1,357	1,345
2000	1,233	1,222	1,212
TOTAL	57,417	56,965	56,573

increase was maintained at 6% per annum. The straight 'inflation' effect would thus not be obscured by an oil price effect of the type investigated before. Two different scenarios were examined, one where the rate of inflation is forecast to be 4% and one where it is assumed to be 12%. The results are presented in Table 4.7. It appears that the time profiles of tax revenues (when deflated to constant prices) are little affected, and consequently overall tax takes are also very similar. A higher rate of inflation does lead to a lower tax take but the effect is small and, perhaps surprisingly, the annual figures do not always reflect the overall conclusion. In 1985,

Table 4.8a The effect on total government tax take of different assumptions regarding the dollar/sterling exchange rate
£ million — constant 1975 prices

Year	*Total government tax take when, from mid – 1982, the average dollar/sterling exchange rate is to remain constant at:*				
	$1.8 – £1	$1.85 – £1	$1.9 – £1 ('*base*' *case*)	$1.95 – £1	$2.0 – £1
1976	20	20	20	20	20
1977	122	122	122	122	122
1978	500	500	500	500	500
1979	995	995	995	995	995
1980	1,781	1,781	1,781	1,781	1,781
1981	2,685	2,685	2,685	2,685	2,685
1982	2,554	2,554	2,554	2,554	2,554
1983	2,742	2,640	2,546	2,457	2,372
1984	3,320	3,208	3,108	3,016	2,969
1985	3,937	3,783	3,614	3,495	3,314
1986	4,209	4,072	3,972	3,814	3,686
1987	4,191	4,046	3,986	3,854	3,733
1988	4,529	4,323	4,122	4,062	3,914
1989	3,654	3,559	3,427	3,240	3,112
1990	3,535	3,337	3,163	3,032	2,941
1991	3,190	3,123	3,009	2,772	2,654
1992	2,893	2,776	2,641	2,659	2,600
1993	2,657	2,577	2,518	2,413	2,319
1994	2,429	2,347	2,271	2,202	2,139
1995	2,254	2,171	2,098	2,030	1,960
1996	2,041	1,975	1,910	1,839	1,767
1997	1,909	1,843	1,781	1,722	1,658
1998	1,684	1,623	1,564	1,510	1,468
1999	1,453	1,403	1,357	1,313	1,270
2000	1,311	1,265	1,222	1,182	1,143
TOTAL	60,594	58,726	56,965	55,269	53,675

for example, the higher the rate of inflation that has been assumed the higher will be the tax take. In general, however, the positive or negative effects which higher or lower inflation rates respectively have on nominal company gross revenues offset—almost exactly—the negative or positive effect which they have on the real value of capital allowances.

Different exchange rate assumptions are presented in Tables 4.8(a) and 4.8(b). The a priori expectation is that, since the price of

Table 4.8b *The effect on total government tax take of different assumptions regarding the dollar/sterling exchange rate*

£ million — constant 1975 prices

Year	*Total government tax take when, from the end of 1982, the average dollar/sterling exchange rate is assumed to:*		
	Remain constant at $1.9 = £1 ('base' case)	*Appreciate (from $1.9 = £1) at 0.5% p.a.*	*Depreciate (from $1.9 = £1) at 0.5% p.a.*
1976	20	20	20
1977	122	122	122
1978	500	500	500
1979	995	995	995
1980	1,781	1,781	1,781
1981	2,685	2,685	2,685
1982	2,554	2,554	2,554
1983	2,546	2,542	2,550
1984	3,108	3,084	3,132
1985	3,614	3,547	3,681
1986	3,972	3,903	4,042
1987	3,986	3,864	4,000
1988	4,122	3,964	4,306
1989	3,427	3,311	3,584
1990	3,163	3,000	3,374
1991	3,009	2,866	3,182
1992	2,641	2,492	2,849
1993	2,518	2,345	2,657
1994	2,271	2,120	2,437
1995	2,098	1,939	2,268
1996	1,910	1,751	2,081
1997	1,781	1,621	1,956
1998	1,564	1,403	1,735
1999	1,357	1,221	1,507
2000	1,222	1,090	1,369
TOTAL	56,965	54,719	59,365

oil is set in dollars, the assumption of a higher exchange rate should yield a lower value of gross revenues and hence government tax take, and vice versa. This is shown to be the case. Constant exchange rate assumptions of £1 = $1.8, $1.85, $1.95 and $2.0 were all simulated and these resulted in increases of 6.4% and 3.1% and decreases of 3.0% and 5.8% respectively in the overall tax take as compared to the 'base' case. The same pattern was also repeated consistently by the annual figures. Moreover, even if the assumption of constancy was relaxed in favour of either an appreciating or a depreciating exchange rate (see Table 4.8(b)), similar results were still forthcoming.

The effect on tax revenues of lower levels of oil depletion is again fairly easy to predict. Table 4.9 gives the results of simulations where oil production is, in one case, 5% and, in the other, 10% below the expected ('base' case) level for all fields after 1981. As expected, lower production results in lower revenues and lower government take. In fact a 5% reduction in the rate of extraction resulted in a 6.1% drop in government revenue, to £53,472 million, while a 10% reduction yielded a 12.5% fall, to £49,858 million. The slightly greater than proportionate fall is due to the fact that petroleum revenue tax is subject to an exempt allowance which is relatively more significant at lower production levels. In addition, an extra simulation was performed where the aggregate production level was assumed to be about 4.5% higher than in the 'base' case. This corresponded to oil extraction of 11.949 million barrels[16] within the period 1976–2000. Total government take was £60,803 million as compared to the £56,965 million for the 'base' case.

In contrast, higher capital and operating costs seem to have little effect on the overall size of government revenues. Table 4.10 gives the results for the two cases where capital and operating expenditure are 5% and 10% respectively above expected values. Even in the latter case, the overall tax take only falls by just over 2%, to £55,608 million. The oil companies appear to shoulder most of the burden.

The simulations discussed so far were all cases where individual assumptions made in the 'base' case varied, and we studied their respective effects on government revenue. There are, of course, many composite variations which could have been tried but our approach was chosen in order to present results which were as clear and easy to interpret as possible. It is, nevertheless, useful to consider in addition a 'pessimistic' case to see how a combination of the individual effects might reinforce one another. One final simulation was thus undertaken where the following changes were made with respect to the 'base' case.

(i) The real sterling price of oil does not increase after 1982.
(ii) The annual rate of oil depletion is 10% lower than expected after 1981.

Table 4.9 The effect on total government tax take of different assumptions regarding the rate of oil depletion

£ million — constant 1975 prices

Year	*Total government tax take when, from the end of 1981, production rates in relation to their expected values are:*			
	10% lower	*5% lower*	*'base case'*	*4.5% higher**
1976	20	20	20	20
1977	122	122	122	122
1978	500	500	500	500
1979	995	995	995	995
1980	1,781	1,781	1,781	1,781
1981	2,685	2,685	2,685	2,685
1982	2,383	2,469	2,554	2,671
1983	2,183	2,363	2,546	2,694
1984	2,752	2,947	3,108	3,546
1985	3,047	3,303	3,614	4,037
1986	3,354	3,674	3,972	4,150
1987	3,397	3,725	3,986	4,409
1988	3,507	3,910	4,122	4,645
1989	2,966	3.102	3,427	4,100
1990	2,656	2,932	3,163	3,773
1991	2,483	2,650	3,009	3,433
1992	2,376	2,588	2,641	3,076
1993	2,117	2,306	2,518	2,832
1994	1,952	2,142	2,271	2,548
1995	1,798	1,958	2,098	2,177
1996	1,637	1,766	1,910	1,856
1997	1,515	1,658	1,781	1,562
1998	1,394	1,467	1,564	1,303
1999	1,175	1,267	1,357	1,023
2000	1,064	1,143	1,222	865
TOTAL	49,858	53,472	56,965	60,803

* See text and reference 16 for details.

(iii) Annual rates of both capital and operating expenditure are 10% higher than expected after 1981.
(iv) The assumptions regarding the rate of inflation and the exchange rate are unchanged.

This scenario is not, of course, the most pessimistic set of assumptions that could have been made. But it will suffice to illustrate the point at issue. The results are presented in Table 4.11. It can be seen that the overall tax take has fallen dramatically to

Table 4.10 The effect on total government tax take of different assumptions regarding capital and operating expenditure

£ million — constant 1975 prices

Year	*Total government tax take when, after 1981, capital and operating costs are above their expected values by:*		
	0% ('base' case)	*5%*	*10%*
1976	20	20	20
1977	122	122	122
1978	500	500	500
1979	995	995	995
1980	1,781	1,781	1,781
1981	2,685	2,685	2,685
1982	2,554	2,552	2,549
1983	2,546	2,523	2,499
1984	3,108	3,077	3,097
1985	3,614	3,556	3,503
1986	3,972	3,966	3,861
1987	3,986	3,923	3,900
1988	4,122	4,139	4,042
1989	3,427	3,318	3,277
1990	3,163	3,096	3,057
1991	3,009	2,821	2,766
1992	2,641	2,721	2,702
1993	2,518	2,458	2,411
1994	2,271	2,248	2,227
1995	2,098	2,071	2,039
1996	1,910	1,875	1,834
1997	1,781	1,749	1,721
1998	1,564	1,537	1,521
1999	1,357	1,340	1,314
2000	1,222	1,204	1,186
TOTAL	56,965	56,276	55,608

£31,637 million, a drop of 45% compared to the 'base' case. Petroleum revenue tax, whose share of the total take was some 57% in the 'base' case, now accounts for less than 48%, reflecting the decline in the level of profits.

4.6 Concluding Remarks

Our stated objectives in this chapter were threefold. The first was to describe the North Sea oil taxation system both as it stood during 1982 and also the subsequent arrangements that were to take effect 1 January 1983. The second was to trace the historical evolution of the system from its initial form, and the third to provide estimates

Table 4.11 Total government tax take from North Sea oil, 1976–2000 ('pessimistic' case—constant 1975 prices)

Year	*Royalties*	*Supplementary Petroleum Duty*	*Petroleum Revenue Tax*	*Corporation Tax*	*Total Government Take*
	(£ million)	*(£ million)*	*(£ million)*	*(£ million)*	*(£ million)*
1976	20	0	0	0	20
1977	116	0	0	6	122
1978	182	0	116	202	500
1979	242	0	402	351	995
1980	411	0	964	406	1,781
1981	496	661	1,089	440	2,685
1982	542	836	553	448	2,379
1983	485	124	1,027	464	2,100
1984	520	0	1,628	725	2,873
1985	541	0	1,117	523	2,180
1986	529	0	1,305	867	2,700
1987	480	0	1,222	680	2,382
1988	415	0	927	551	1,893
1989	353	0	960	493	1,806
1990	302	0	812	392	1,505
1991	259	0	582	311	1,153
1992	220	0	474	291	985
1993	190	0	409	232	832
1994	166	0	341	178	685
1995	143	0	278	129	550
1996	124	0	222	85	430
1997	109	0	195	62	366
1998	90	0	160	40	291
1999	73	0	141	27	241
2000	62	0	105	17	184
TOTAL	7,069	1,621	15,027	7,920	31,637

of the likely magnitude of government tax revenues over the period from 1976 to 2000. The first two objectives were essentially descriptive and hence presented no real problems of content. To fulfil the third objective, it was decided that the best approach was first to provide revenue forecasts using a set of 'base' case assumptions regarding production profiles, capital and operating costs, the price of oil, inflation and exchange rates etc. Then each of these assumptions was in turn subjected to scrutiny to see how any variations might affect the overall results. Simulations were thus undertaken for a variety of cases where each of the assumptions were individually changed.

The conclusions of this exercise were that the tax estimates were most sensitive to changes in the rate of depletion and in the real sterling price of oil—whether the latter was effected through changes in the nominal dollar price or via movements in the exchange rate. The tax estimates were largely insensitive to different levels of capital and operating expenditure. Different assumptions related to the UK inflation rate also seemed to have little effect on taxation revenue, as long as the nominal oil price was allowed to rise concomitantly. And lastly, a simulation was carried out on a 'pessimistic' case where the sensitivity of the calculations to changes in the variables was examined.

5 Quantification of the Model Parameters and Projection of the Exogenous Variables

5.1 Introduction

Thus far in this study we have developed a long-run macroeconomic model characterizing an industrialized economy with a flexible exchange rate and which is endowed with an exhaustible resource. The model is developed with special reference to the UK economy. We shall now show how the analytical solutions reached formally can be verified numerically.

In order to apply the model and obtain numerical values for the optimal trajectories of investment and oil depletion, we need to supply it with a certain amount of initial data. Such data comprise the values of the parameters of the model and the constraints, the projected time path of the exogenous variables and the values of the state variables in the starting year.

This chapter is devoted to the preparation of the data for the utilization of the model. It should be made clear at the outset that the values of the model parameters are quantified by means of economic reasoning. They are not determined through econometric estimation techniques. This is because it is felt that the experience of the past may not be necessarily relevant to the future, when the emergence of the oil sector can of itself generate economic changes which were not present in the pre-oil period. For instance, the mere presence of this newly-developed resource created a certain confidence in the UK economy within the international currency market which culminated in a rise in the value of sterling in the late 1970s.

Using parameter values which are the result of applying data related to the pre-oil period may not be suitable to explain future behaviour. Thus it was decided to assign numerical values to the parameters by studying the underlying economic theory and outlining the rationale for selecting each set of parameter values, even though this proved much more difficult in practice than the straightforward econometric estimation method. Wherever possible, of course, reference is made to the existing literature and to some of the econometric results in other studies.

There may be some objection regarding the values of some of the parameters quantified in this chapter. Indeed one could claim that changes in these parameter values may affect the numerical solutions of the model. The answer is that if any of the parameter values prove to be an underestimate or an overestimate in reality, the effect on the overall solutions can be discovered by carrying out some sensitivity analysis. This means we can vary the size of each parameter by both raising it and reducing it.

Finally, it should be pointed out that another reason for adopting this method of estimation is related to the nature of our study, which is primarily theoretical. This research is designed to demonstrate a methodology in planning, applying optimal control theory. Thus it is important to emphasize that it is not so much the absolute values of the solutions that matter. Rather the relative size of the figures should be used to understand the implications of the solutions.

5.2 Domestic Non-Oil Production Function

The problems inherent in the use and estimation of aggregate production functions are substantial[1]. Nevertheless, it is important to establish the relationship between what is put into a productive process and what emerges in the form of output. In introducing a production function in a macromodel, the question of the appropriate functional form arises. There are a number of possible alternatives of which the input-output, the Cobb-Douglas, the constant elasticity of substitution and the transcendental logarithmic functions[2] are perhaps the most common.

In selecting the production function it was important to apply a non-linear function that would readily lend itself to an analysis of returns to scale. This characteristic was crucial in this exercise, since the parameters of the function were to be quantified on the basis of economic arguments. With these considerations in mind, the following functional form is postulated:

$$D_t = g(1+\beta)^t K_{t-1}{}^{\alpha} Lt_{-1}{}^{\gamma}$$

where D_t = Non-oil output
K_t = Capital input
L_t = Labour input
t = time

The parameters are defined as follows:

α = Exponent of capital
γ = Exponent of labour
β = Rate of technical progress
g = Scale factor

Suffice it to say that each of the variables in the above equation are very difficult to measure. Indeed, aggregating dissimilar quantities is a notoriously difficult task, and problems always arise with their measurement. Conceptually, on the input side we are interested in measuring the amount of human effort, that is, labour, applied directly to production and the size of capital stock used in the production process. On the output side, our interest is in the volume of output produced at the end of the production process. The following specific definitions are used for each of the variables of the production function.

5.2.1 Measurement of variables

(i) *Non-oil gross domestic product*

The measure chosen to represent output is gross non-oil domestic product at factor cost. As Rita Maurice notes, 'Valuations at factor cost displays the composition of national product in terms of factors of production employed'.[3] Output is thus measured net of any taxes (or subsidies) on expenditure levied by the government.

(ii) *Non-oil gross domestic capital stock*

The theoretical difficulties of measuring capital stock[4] for use in production functions have been well documented elsewhere[5] and should not detain us here. What is of importance is which of the available published measures is most appropriate in the present context, since there is no uniquely correct value of fixed assets.

The decision with respect to choosing a net or a gross measure of fixed assets was made on the basis of uniformity between the variables capital and output. So long as the two variables were expressed in the same form both gross or both net of capital consumption, consistency would be ensured within the production function. For our model we chose the gross measure. As Tom Griffen remarks, 'For an indicator of the current gross output potential of fixed assets the more suitable series is gross capital stock in preference to net capital stock'.[6] The specific measure of "capital" input which has been adopted here is: gross domestic non-oil capital stock at 1975 replacement cost.

(iii) *Labour*

With regard to the 'labour' input, one method is to measure the amount of time spent at the place of work (in man-hours). Clearly, however, an hour of different people's time is not necessarily equivalent in terms of their labour services. One approach to overcoming this drawback is to weight different types of labour by their corresponding wage rates in an attempt to reflect relative efforts. While this would be an interesting exercise, it would be beyond the scope of this study. A more readily accessible measure is that given in the *National Income and Expenditure* book on the total employed labour force in the non-oil sector.[7] This includes all labour except that employed in the petroleum and natural gas industries. The labour input for the years 1976–80, together with projected figures for the years 1980–2000, are shown in Table 5.13 at the end of this chapter.

5.2.2 *Measurement of parameters*

We now come to the task of assigning estimates to the parameters of the production function itself. In view of the multitude of problems with which econometric estimation would have been beset—for instance, specification of the error term, choice of either a single or a system of equations to describe the production model, aggregation bias, etc.—it was decided to assign values to the parameters on the basis of economic reasoning instead.

Initially, the rate of technical progress has been taken to be 0.012 though this value is something which will certainly be varied in chapter 7 where sensitivity analyses are carried out. Hence,

$$\beta = .012.$$

We begin with the exponents of the capital stock, α, which also represents the elasticity of response of the capital input in the production function. It was felt that the value of α should be chosen such that the rate of return generated on domestic investment should approximate to the most suitable measure of profitability normally quoted by the Department of Trade and Industry.[8] Both the two sets of figures given by the Department of Trade and Industry—the net and the gross rates of return—measure accounting rates of return (that is, the return achieved in a particular year on total capital employed on all projects then in existence), and so are, in that sense, conceptually very similar to our own expression. The gross rate was adopted here, since K_t is defined in the model as gross stock of fixed assets. The relevant data for the years 1976–9 is shown in Table 5.1. In the model the incremental rate of return on domestic investment is generated by differentiating output with respect to the capital stock at the end of the previous year. Thus,

$$\frac{\partial D_t}{\partial K_{t-1}} = \alpha g(1+\beta)^t K_{t-1}^{\alpha-1} L_{t-1}^{\gamma}$$

$$= \alpha \frac{D_t}{K_{t-1}}.$$

Therefore, an appropriate value for the marginal productivity of capital can be chosen by comparing the rates of return provided in Table 5.1 with the output-capital ratios shown in table 5.2. A comparison of this kind suggests that a value of 0.32 for the exponent of capital would be reasonable. This means:

$$\alpha = 0.32$$

Such a value for α would generate a rate of return on domestic capital that would be, on average, about 5%.
The exponents of labour should be quantified ideally such that the marginal product of labour is equal to the going wage rate. While it is possible to arrive at an aggregate measure of profitability to represent all the investments in the economy, it would be difficult to arrive at a representative wage rate for the country. This is because, as mentioned above, the labour force is not homogeneous. Thus the exponent of this variable was calculated in conjunction with the scale variable in such a way that when the values of K_t and

Table 5.1 Companies' rate of return on capital employed 1976–1979

Year	*Gross rate of return of Industrial and commercial Companies excluding North Sea production and Exploration activities (%)*	*Gross rate of return of Manufacturing companies (%)*
1976	6.3	5.0
1977	6.9	6.1
1978	7.0	6.2
1979	5.9	5.0

Source: British Business, Vol. 3, No. 5, (3 October 1980), pp. 222 - 3.

Table 5.2 Output—Capital Ratios

Year	*Non - oil GDP at factor cost 1975 prices (£ million)*	*Gross non - oil domestic Capital stock at 1975 Replacement cost (£ million)*	*Output - Capital Ratio*
1976	97,551	470,000	0.207
1977	97,536	482,800	0.202
1978	100,214	494,800	0.203
1979	99,402	506,900	0.196

Source: National Income and Expenditure 1980, edition (London, HMSO 1980), pp. 3, 12 and 84.

L_t are plugged into the production function it would generate values of D_t that would come closest to the actual value of th UK non-oil GDP for the first four years of our planning period, 1976–80.

To sum up, the following parameter values were assigned:

$$\gamma = 0.8$$
$$g = 112.33$$

5.3 Raw Material Usage in the Economy

In equation 2.3 of the model, raw material usage is postulated as a constant proportion, ξ, of domestic output:

$$R_t = \xi D_t$$

An estimate is thus needed for the parameter ξ. The data shown in table 5.3 has been gathered from the most recent Input-Output tables available for the years from 1970 to 1974. As can be seen, the ratio of raw material inputs to gross domestic product changes sharply in 1974 as the value of imports increases dramatically after the first round of OPEC oil price rises. Unfortunately data are not available for the years after 1974, but it is to be expected that the United Kingdom would have adjusted to the higher costs of raw materials by raising the price of output. Hence it was thought that the ratio of raw material input to gross domestic product would fall again in subsequent years from its value of 112.9% in 1974.
It should be pointed out that although in our definitions,[9] D_t is defined as the non-oil output, here we use GDP as our proxy variable, because North Sea oil had not come into effective operation in the years when the input-output statistics were compiled[10]. Therefore, the ratio of total aggregate raw materials to GDP can be used here to represent the ratio of R_t to D_t.

Given the range of figures in the last column of Table 5.3 an average ratio of 95% seemed a plausible assumption in the first instance. Later, one could simulate against variations around this parameter. Thus:

$$\xi = 0.95$$

Table 5.3 Analysis of raw material inputs to UK domestic production

Year	*Domestic input of commodities*	*Input of Imported goods and services*	*Total input of raw Materials*	*Gross domestic Product at Factor cost*	*Raw materials as proportion of GDP*
	(£ million)	*(£ million)*	*(£ million)*	*(£ million)*	*(%)*
	(1)	*(2)*	*(3)*	*(4)*	*(5)*
1970	34,238.8	7,465.4	41,704.2	42,778	97.5
1971	37,107.6	7,952.2	45,059.8	48,432	93.0
1972	39,675.8	8,649.6	48,325.4	54,679	88.4
1973	65,889.7	18,143.0	84,032.7	74,414	112.9

Notes: (i) All monetary quantities in current prices.
(ii) The figures quoted for 1970 are consistent with the national accounts for that year in *National Income and Expenditure, 1973 edition.* For 1971, 1972 and 1974, the corresponding issues of the Blue Book are *National Income and Expenditure 1963–1973, National Income and Expenditure 1964–1974* and *National Income and Expenditure, 1980 edition,* respectively.

Sources: 1970 : *Input-Output Tables for the United Kingdom 1970*, Business Monitor PA1004 (London: HMSO, 1974) p.17.
1971 : *Input-Output Tables for the United Kingdom 1971*, Business Monitor PA1004 (London: HMSO, 1975) p.19.
1972 : *Input-Output Tables for the United Kingdom 1972*, Business Monitor PA1004 (London: HMSO, 1970) p.19.
1974 : *Input-Output Tables for the United Kingdom 1974*, Business Monitor PA1004 (London: HMSO, 1981) p.23.

5.4 The External Income

The stock of overseas assets is a concept which is very hard to define unambiguously. In the model, we define it as the cumulative sum of net investment overseas, where the latter is set equal to the current balance in the balance of payment as shown in equation 2.11 of the model.[11] This definition is valid because the macroeconomic model is formulated such that it shows the working of the economy in the long run, and the current account balances with the capital account over a long period.

Following R. Maurice we take investment abroad to represent the net increase in the value of overseas assets acquired by the UK residents less the net income in the value of assets in the United Kingdom acquired by non-residents.[12] Overseas assets thus include investment in physical assets, financial assets, gold and convertible currency reserves.

It has been postulated in the model equation 2.8[13] that net property income from abroad is a lagged, linear function of the stock of overseas assets, namely

$$F_t = \theta E_{t-1}$$

where E_{t-1} = stock of overseas assets in the previous period

F_t = net property income from abroad.

Again this formulation, as envisaged, represents the behaviour in the long run. Since the equation is an aggregate one, it has not been thought necessary, in the context of this study, to disaggregate further and try to enter the complex relationship between investment expenditures and the time patterns of income flows which are thereby generated. Nevertheless, it was felt that our formulation captured the essence of the relationship we were trying to model and would therefore suffice for the purposes of this exercise. One corollary of this problem was that it was not possible to assign a value to the parameter θ on the basis of calculated ratios from past data of net property income and asset stock. Instead, it was assumed that the rate of return on overseas assets would be equal to the real interest rate available. Samuel Brittan[14] provides the estimates in Table 5.4 of the real interest rate for France, the United States, Germany and Japan. A value of 7% has thus been chosen in the first instance to represent the real yield on overseas assets. This implies,

$$\theta = 0.7$$

Table 5.4 *Estimation of the real interest rate in selected countries abroad*

Country	*Interest rate (3 months two rate, rounded)* (%)	*Inflation rate (consumer prices rise in year to July)* (%)	*Approx. real Interest rate* (%)
France	21-22	13.4	8
USA	17-17½	10.7	7
Germany	11½-12	5.8	6
Japan	7¼-7½	4.3	3

Source: Financial Times, (17 September, 1981), p.23.

5.5 The Consumption Function

The formulation of the consumption function adopted in the macromodel is consonant with the results of many empirical studies of past consumer behaviour. Two major examples are Simon Kuznet's fundamental contribution on consumption and saving behaviour,[15] and the seminal work of Milton Friedman's Permanent Income Hypothesis.[16] One of the important conclusions of Kuznet's work[17] is that over the long run the value of the marginal propensity to consume becomes on average, close to that of the average propensity to consume, as income grows along a trend. Similarly Milton Friedman found, using USA data, that his best estimate of the Permanent Income Hypothesis gave an insignificant intercept for the consumption function. This implied that in the long run the marginal and the average propensities to consume converge.[18] Friedman used both cross-section and time-series data.

Since the problem which we are addressing in this study is also long-run, it was considered appropriate to postulate a functional form in which the consumption function passed through the origin. The particular form of the function which was adopted was

$$C_t = \eta\ (D_t + F_t - l_t^{111})$$

where

C_t = Consumers' expenditure
D_t = Gross non-oil domestic product at factor cost
F_t = Net property income from abroad
l_t^{111} = Taxes on income plus national insurance contributions
The term $(D_t + F_t - l_t^{111})$ is basically equivalent to disposal income.

A value has now to be assigned to the parameter, and it is instructive to examine some of the estimates provided in the relevant empirical literature on the UK economy.

The Cambridge Growth Project use a simple consumption function of the Absolute Income type in their condensed model.[19] Moreover, they provide estimates for both linear and log-linear functional forms, the results of which are shown below:

$$C_t = 63.5 + 0.7302\ Y_t$$
$$\text{In } C_t = 2.069 + 0.7933 \text{ In } Y_t$$
$$\text{where } C_t = \text{Consumers' expenditure}$$
$$Y_t = \text{Personal disposal income}$$

The marginal propensity to consume out of real personal disposable income in the Growth model is 0.73, and the expenditure elasticity is 0.79—that is, a 10% increase in real income gives only an 8% increase in real consumer spending.

The other major modelling concerns on the UK economy adopt approaches from which detailed comparisons of estimates are not possible. Moreover, their studies are concerned primarily with short-run consumption functions,[20] as opposed to long-run, which is the case in our model. We should, perhaps, mention as an example of a completely different methodology, the work undertaken by the Cambridge Economic Policy Group.[21] They estimate an aggregate equation in which consumption plus non-North Sea investment are expressed as a function of current and lagged income, hire purchase debt and interest rates. An investment equation is then estimated in terms of disposable wages and salaries, disposable property income and an interest rate term. Finally, the consumption function is obtained by deducting investment from the aggregate relation.

Given the results of these various studies and bearing in mind the difficulties associated with trying to summarize a complex process, it was decided in the first instance to attach a value of 83% to the parameter representing the average/marginal propensity to consume. This is slightly higher than the value of the marginal propensity to consume estimated by the Cambridge Growth Project, which was 0.73, as shown earlier. The reason for opting for a slightly higher value is because we are assuming away the intercept term which appears in their equation. Thus to begin with we assign:

$$\eta = 0.83$$

5.6 Taxation

(i) *Income-related tax*

Taxes on income and national insurance contributions have been consolidated in the model to form the category of income-related taxes. Such taxes are expressed in equation 2.12 of the macro model as a linear function of gross non-oil national product at factor cost, that is,

$$l_t^{111} = \varepsilon(D_t + F_t)$$

where D_t = Gross non-oil domestic product at factor cost
F_t = Net property income from abroad
l_t^{111} = Taxes on income plus national insurance contributions

Table 5.5 has been constructed using data from the past decade and, as can be seen from the final column, over this period

Table 5.5 Derivation of the estimate of the rate of income related tax

Year	*Gross non-oil Domestic product at factor cost (£ million) (1)*	*Net property Income from Abroad (£ million) (2)*	*Gross non-oil National product at factor cost (£ million) (3)*	*Income-related Tax payments (£ million) (4)*	*Income-related Tax payments as proportion of non-oil GNP (%) (5)*
1970	43,504	554	44,058	10,043	22.79
1971	49,453	502	49,955	10,829	21.68
1972	55,231	538	55,769	11,437	20.51
1973	64,215	1,268	65,483	13,101	20.01
1974	74,409	1,428	75,837	17,584	23.19
1975	93,970	771	97,471	23,506	24.12
1976	110,728	1,305	112,033	27,318	24.38
1977	124,250	46	124,296	29,890	24.05
1978	141,946	520	142,466	32,552	22.85
1979	158,536	289	158,825	36,613	23.05

Notes: (i) All monetary quantities are expressed in current prices.

Sources: (1) *National Income and Expenditure: 1980 Edition* (London: HMSO, 1980) p.12.
(2) *National Income and Expenditure: 1980 Edition* (London: HMSO, 1980) p.3.
(3) *National Income and Expenditure: 1980 Edition* (London: HMSO, 1980) p.58.

income-related taxes have varied between 20% and 25% as a proportion of non-oil gross national product. The tax totals, however, include tax receipts from the North Sea sector which are to be treated as a separate entity in this analysis. It was therefore felt that a value at the bottom end of the observed range should be adopted, and a figure of 20% was assigned to the parameter ε as a first approximation. Thus,

$$\varepsilon = 0.20$$

(ii) *Expenditure tax*
Taxes, less subsidies, on expenditure are specified in the model, equation 2.12 as a simple proportion of consumers' expenditure at market prices. Thus,

$$l_t^{11} = \sigma C_t$$

where C_t = Consumers' expenditure
l_t^{11} = Taxes on expenditure less subsidies

Table 5.6 contains the necessary information for arriving at the estimate of the parameter σ. In a similar fashion to income-related taxes, expenditure tax has risen through the 1970s, but here a value

Table 5.6 Derivation of the estimate of the rate of expenditure tax

Year	*Consumers' expend - iture at 1975 market Prices (£ million) (1)*	*Taxes (less subsidies) on expediture at 1975 prices (£ million) (2)*	*Taxes (less subsidies) as proportion of consumers' Expenditure (%) (3)*
1970	57,814	9.107	15.86
1971	59,724	9,514	15.93
1972	63,270	10,243	16.19
1973	66,332	10.918	16.46
1974	64,881	10,638	16.40
1975	64,824	10,459	16.13
1976	64,642	10,823	16.74
1977	64,240	10,866	16.91
1978	68,074	11,919	17.51
1979	70,816	12,349	17.44

Notes: All monetary quantities are at constant 1975 prices.
Sources: (1), (2) *National Income and Expenditure: 1980 edition.* (London, HMSO 1980) p. 17.

of 17% is assigned—towards the top end of the range and more relevant to the second half of the decade, so that,

$$\sigma = 0.17$$

(iii) *Petroleum tax*

As has already been mentioned, the tax receipts from the North Sea oil sector are treated separately, given the importance of this revenue *vis-à-vis* the planning problem posed in this study. Details of the calculations for this tax were explained in chapter 4.

5.7 Exports

Total exports in the model have been disaggregated into two components in order to highlight the particular role of the oil sector. The following equation summarizes the relationship between the chosen aggregates, where:

$$X_t = V_t + \omega P_t$$

X_t = Total exports
V_t = Non-oil exports
P_t = Oil revenues
ω_t = Proportion of oil exported

(i) *Oil exports*

The term (ωP_t) is, hence, the amount of indigenous oil production which is exported. An estimate for the value of the parameter ω was obtained using data for the years 1976—9 shown in Table 5.7. As can be seen, the percentage of oil exported rose steadily over the first few years of domestic oil production although the upward trend may well have levelled off to yield a steady proportion of about 50%. This is the figure we have adopted as the long-run percentage of North Sea oil that is exported. Thus,

$$\omega = 0.50$$

(ii) *Non-oil exports*

Non-oil exports have been assumed to grow along a long-run trend but to vary according to the level of the exchange rate. Hence,

$$V_t = m_0 (1 + \nabla)^t S_{t-1} m_1 \tag{5.1}$$

where S_t = Real sterling effective exchange rate index (1975 = 1.0)
m_0 = Scale parameter
m_1 = Price elasticity of demand for exports
∇ = Trend rate of growth of non-oil exports per annum
t = time

The trend has been included to try and capture the effects of such

Table 5.7 Analysis of shipments of oil exports

Year	Shipments abroad of indigenous Oil (Thousand tonnes)	Total indigenous Oil production (Thousand tonnes)	Proportion of Oil exported (%)
1975	-	1,567	-
1976	2,935	12,171	24.11
1977	15,611	38,265	40.80
1978	23,948	54,006	44.34
1979	39,044	77,854	50.15

Sources: 1975 *Energy Trends* (London: HMSO, August 1976) p. 7.
1976 - 79 *Energy Trends* (London: HMSO, July 1980) p. 7.

variables as the 'state of world trade' which are normally present in export equations in forecasting models but which must necessarily be excluded in a growth model of this type. The value of 2% per annum chosen for the trend growth rate reflected the approximate increase in export volume (net of exchange rate effects) over the decade of the 1970s. As for the parameters which govern the effect of movements in the exchange rate, a preliminary assumption was made that the relationship would be strictly inverse—i.e., the value of the response parameter, m_1, was set equal to -1. The level of non-oil exports in 1976 can be calculated as follows:

Exports of goods and services*	£29,494 m	(at 1975 prices)
Exports of oil†	£133 m	(at 1975 prices)
Non-oil exports	£29,361 m	

Sources: **National Income and Expenditure, 1980 Edition* (London: HMSO 1980) p.17.
†Calculated from the data given in Tables 5.7 and 5.13 using a conversion factor of 7.5 barrels = 1 tonne.

Now, as the index of the real sterling effective exchange rate has been taken to equal 1.0 in 1975, this implies a derived value for the parameter, m_0, of 29,361. However, since it would be misleading to accept such a degree of precision from this estimation procedure, a round figure of 30,000 has been adopted for the parameter m_0. The values assigned to the parameters of the non-oil export equation can be summarized thus:

$$m_0 = 30\ 000$$
$$m_1 = -1$$
$$\nabla = 0.02$$

5.8 The Exchange Rate Equation

Besides its substantial contribution to government revenues, perhaps the most important way in which North Sea oil will have an influence on the rest of the British economy will be through its effect on the exchange rate. Indeed, modelling exchange rate movements is by no means simple. Any attempts at precise forecasting of short-term fluctuations would necessitate, amongst other things, a knowledge of relative interest rates, an understanding of international capital flows and an appreciation of changing expectations. However, it is worth stressing that our problem here is essentially long-run and therefore, to include a multitude of explanatory variables in the exchange rate equation is not necessary, even if it is possible. Our focus of attention is specifically the effect of North Sea oil production on the variables S_t. Therefore, we have tried to capture the direct effect of the oil revenue on the exchange rate by postulating the non-linear relationship shown below, which appears as equation 2.17 of the model.[22] It should be noted that we have chosen to use this functional form to generate values for the real sterling effective exchange rate, the actual values of which for the years 1973—80 are given in Table 5.8.

$$S_t = \Omega_0 + \Omega_1(\pi_t\, q_t)\Omega_2$$

where S_t = Real sterling effective exchange rate index (1975 = 1.0)
π_t = Real sterling price of oil
q_t = Rate of oil extraction
Ω_0 = Effective exchange rate in absence of North Sea Oil
Ω_1, Ω_2 = Response parameters

The assignment of realistic parameter values is particularly difficult. Having mentioned some of the variables that would have to be included in any forecasting equation, it is obviously unjustifiable to use actual data to generate the estimates. Instead rough orders of magnitude have to be estimated intuitively. Hence the 'no-oil' exchange rate, Ω_0, was taken to be 0.875. As can be seen from Table 5.8, this estimate is lower than the actual values derived for the real sterling effective exchange rate index for the years 1973 to 1980. However, in as much that such values were possibly inflated due to the expectation of future oil revenues, a low value of the 'no-oil' exchange rate was deemed appropriate.

The parameters that show the response of the exchange rate to the emergence of the oil income are Ω_1 and Ω_2. The exponent Ω_2 captures the impact of changes in oil revenue. Given the nature of the oil sector, and notwithstanding that its emergence generated

Table 5.8 Derivation of the real sterling effective exchange rate index

Year	*Sterling effective Exchange rate index (average 1975=1.0)* (1)	*UK consumer Price index* *(1975=100)* (2)	*Industrial countries consumer price index* *(1975=100)* (3)	*Real sterling Effective exchange rate index (1975=1.0)*
1973	1.114	69.4	79.5	0.972
1974	1.083	80.5	90.0	0.969
1975	1.000	100.0	100.0	1.000
1976	0.857	116.5	108.3	0.922
1977	0.812	135.0	117.5	0.933
1978	0.815	146.2	125.9	0.946
1979	0.873	165.8	137.4	1.053
1980	0.961	195.0	153.8	1.222

Sources: (1) *United Kingdom Balance of Payments, 1981 edition*, (London: HMSO, 1981) p. 77.
(2) *International Financial Statistics, 1981 Yearbook* (Washington: IMF, 1981) p. 437.
(3) *International Financial Statistics, 1981 Yearbook* (Washington: IMF, 1981) p. 65.

confidence in the pound and boosted its value, its influence is ultimately limited. That is to say, as oil revenue rises, so the impact of the additional revenue begins to fall. Thus the value of Ω_2 would lie between zero and unity. The experience of the late 1970s and early 1980s indicates that the value of Ω_2 is closer to the lower bound. This led to our choosing a value of 0.30 for the exponent of $\pi_t q_t$. Since Ω_1 acts as the scale variable, the corresponding value of it was estimated at 0.0075. The combination of these parameters generated real exchange rate values that came closest to those of the late 1970s and early 1980s. To summarize then:

$$\Omega_0 = 0.875$$
$$\Omega_1 = 0.0075$$
$$\Omega_2 = 0.30$$

5.9 The State Constraint

The system of equations constituting the macroeconomic model is closed by the specification of the state constraint which is shown below.

$$M_t \geq \phi R_t + \mu C_t$$

where M_t = Imports
R_t = 'Raw material' input to domestic production
C_t = Consumers expenditure
ϕ = Proportion of raw materials imported
μ = Proportion of consumption goods imported

Such a formulation is necessary to avoid the obviously unrealistic result of imports taking unduly small (even negative) values—an eventuality which might have arisen as the computer algorithm attempted to maximize the current account surplus, and hence the level of overseas investment. The minimum normal requirement of 'essential' imports which is specified here is assumed to consist of two components: an 'essential' level of raw material imports and an 'essential' level of consumer goods imports. To derive realistic estimates of the two parameters, ϕ and μ, it is necessary once again to consult the *Input-Output Tables*. Unfortunately, this means that data for the past decade are only available for the years 1970, 1971, 1972 and 1974. Table 5.9 gives the relevant figures for the calculation of the value of ϕ, the proportion of the raw materials that are imported. The high value of 21.59% in 1974 probably reflects the sharp oil price increases of that period. (See Figure 5.1 for a graphical illustration of the oil-price changes during the 1970s. If we assume on the one hand that such price rises eventually feed through to the prices of other materials and, on the other hand, that Britain's requirement of imported oil declines as domestic production starts, a value of 0.18 for the parameter of ϕ seems plausible.

Table 5.9 Analysis of the import requirement of raw materials

Year	*Input of Imported goods and services to Domestic production (£ million)*	*Total input of raw materials domestic production (£ million)*	*Proportion of raw materials Imported (%)*
1970	7,465.4	41,704.2	17.90
1971	7,952.2	45,059.8	17.65
1972	8,649.6	48,325.4	17.90
1974	18,143.0	84,032.7	21.59

Notes: (i) All monetary quantities are expressed in current prices
(ii) See Table 5.3 for sources.

Figure 5.1 Price of Arabian light crude oil, 1970—1981

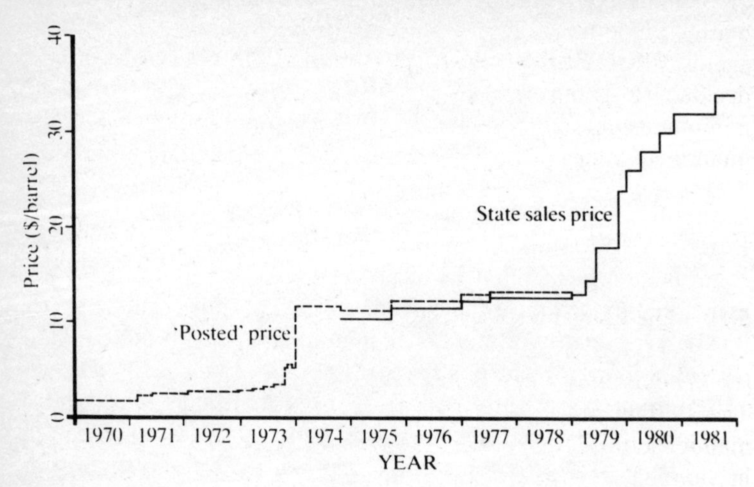

With regard to consumer goods, the results presented in Table 5.10 demonstrate that the proportion of final consumer demand satisfied by imports steadily rises over the five-year span. In the absence of more recent figures, it is not possible to ascertain whether this upward trend has persisted into the latter half of the decade. However, it was thought unlikely both that this degree of import penetration might have fallen significantly of its own accord

Table 5.10 Analysis of the Import Requirement of Consumer Goods

Year	*Consumers' Final demand for imports (£ million)*	*Total Consumer Demand (£ million)*	*Proportion of Consumer goods Imported (%)*
1970	2,183.8	31,404	6.95
1971	2,579.6	34,881	7.40
1972	3,187.0	39,635	8.04
1974	5,127.8	52,489	9.77

Notes: (i) All monetary quantities are expressed in current prices
Sources: 1970: *Input - Output Tables for the United Kingdom 1970*, Business Monitor PA1004 (London: HMSO, 1974) p. 17.
1971: *Input - Output Tables for the United Kingdom 1971*, Business Monitor PA1004 (London: HMSO, 1975) p. 19.
1972: *Input - Output Tables for the United Kingdom 1972*, Business Monitor PA1004 (London: HMSO, 1970) p. 19.
1974: *Input - Output Tables for the United Kingdom 1974*, Business Monitor PA1004 (London: HMSO, 1981) p. 23.

or that the government would have the ability to reduce it as a matter of policy (for fear of reprisals by overseas governments against UK imports). Given these assumptions, it was eventually decided to assign a value of 10% to the parameter, μ, as the minimum import requirement of consumer goods. Thus, the quantified values of the parameters of the state constraints are:

$$\phi = 0.18$$
$$\mu = 0.10$$

5.10 The Exogenous Variables

(i) *The Sterling Price Path of Oil*
The sterling price path of oil has been derived in the following manner. For the years 1976—80, the annual average dollar oil price in nominal terms has been taken as that quoted by Wood, Mackenzie & Co. for Forties marker crude. (Details of these calculations are shown in Table 5.11). The corresponding real 1975 prices were obtained by converting the dollar prices to sterling and then deflating using the wholesale price index for manufactured products. For the years 1979 and 1980, although a similar calculation procedure was adopted, quarterly data (instead of annual data) were used, and the final annual value quoted for the real oil price was derived as an average of the quarterly figures. This is demonstrated in Table 5.12.

For the remaining years under consideration (1981-2000), a certain annual rate of growth of the sterling price of oil in real terms had to be projected. Such a growth rate would reflect some form of long-term average. Needless to say, the price of oil can be extremely unstable so that whatever projection is used, it may turn out to deviate considerably from reality.

The volatile nature of the oil market can radically change expectations of the future within a short time. The case in point is the sharp drop in the prices of oil in early 1982, whereby spot prices sank as low as $26 a barrel. Anticipation concerning the price of oil changed dramatically from a previous expected average rise of 4%-8% in real terms for the 1980s decade to one of acute pessimism. Oil industry experts forecast a real fall in the first half of the 1980s, the price only reaching its 1981 level in real terms by 1990. While the prediction of a fall in real terms in the first half of the eighties decade would be justifiable, what happens in the second half of this decade is subject to a great deal of uncertainty.

To understand the causes of uncertainty in the future price of oil, it is worthwhile examining why the price fell in 1981/82. The factors responsible were the economic recession in the industrialized countries (although the recession was world wide), energy conservation, substitution of other forms of fuel for oil and finally, destocking of oil. The last variable is an important factor in grasping the behaviour of the world petroleum market. Changes in oil stocks act as a cumulative force upon fluctuations in the price of oil. When the price rises, stockpiling pushes the price still higher; when the price falls destocking creates a bigger glut and forces the price down still further. This last factor was partly responsible for the plummeting of petroleum prices in the first quarter of 1982. By the same token, since stocks were run down in the early 1980s, tightening of the market could force the price of oil up because of the expected stock piling that might accompany it.

Table 5.11 Derivation of the real price path of oil

Year (quarter)	*Average oil Price in Nominal terms ($/barrel)* (1)	*Average £/$ Exchange rate* (2)	*Wholesale Price Index (1975=100)* (3)	*Annual average oil price in Real 1975 prices (£/barrel)* (4)
1976	12.80	1.8049	117.3	6.05
1977	14.00	1.7455	140.4	5.71
1978	13.80	1.9197	153.3	4.69
1979	20.67	2.1221	171.9	5.63
1980	34.99	2.3281	200.0	7.51

Sources: (1) 1976 - 1978 Wood, Mackenzie & Co., *North Sea Report*, no. 74 (20 June 1979) p. 3.
1979 (I-IV) Wood, Mackenzie & Co., *North Sea Report*, no. 81 (24 January 1980) p. 39. Monthly figures quoted.
1980 Wood, Mackenzie & Co., *North Sea Report*, no. 93 (28 January 1981) p. 7. Monthly figures quoted.

(2) 1976 - 1979 Central Statistical Office, *Financial Statistics*, no. 215. (London: HMSO, March 1980) p. 130. Monthly figures.
1980 Central Statistical Office, *Financial Statistics*, no. 227 (London: HMSO, March 1981) p. 26. Monthly figures quoted.

(3) Central Statistical Office, *Economic Trends*, no. 317. (London: HMSO, March 1980) p. 42.
1980 (I-IV) Central Statistical Office, *Economic Trends*, no. 329 (London: HMSO, March 1981). p. 42.

One could also look at the past price behaviour of crude oil. Figure 5.1 shows the price path of Saudi Arabian light marker crude over the period 1970 – 81 to which the prices of the various grades of North Sea oil were related during that period. (Of course, there was a sharp deviation from this in March 1982 when the British National Oil Corporation announced a price of $31 while the agreed

Table 5.12 Comparison of the estimated and actual out-turn values for the real price of oil in 1980

Quarter	*Average oil Price in Nominal terms ($/barrel)*	*Average £/$ Exchange rate*	*Wholesale Price Index (1975=100)*	*Quarterly average oil price in Real 1975 prices (£/barrel)*	*Annual average oil price in Real 1975 prices (£/barrel)*
	(1)	*(2)*	*(3)*	*(4)*	*(5)*
1979 (I)	15.50	2.0160	161.6	4.76	
1979 (II)	19.09	2.0806	168.0	5.46	5.63
1979 (III)	23.20	2.2331	176.4	5.89	
1979 (IV)	25.07	2.1588	181.8	6.39	
1980 (I)	32.26	2.2547	191.4	7.48	
1980 (II)	35.17	2.2862	199.0	7.73	7.51
1980 (III)	36.25	2.3819	203.6	7.48	
1980 (IV)	36.25	2.3872	206.1	7.37	

Sources: For 1979 (1) 1976 - 1978 Wood, Mackenzie & Co., *North Sea Report*, no. 74 (20 June 1979) p. 3.
1979 (I-II) Wood, Mackenzie & Co., *North Sea Report*, no. 81 (24 January 1980) p. 39. Monthly figures quoted.

(2) Central Statistical Office, *Financial Statistics*, no. 215. (London: HMSO, March 1980) p. 130. Monthly figures quoted.

(3) Central Statistical Office, *Economic Trends*, no. 317. (London: HMSO, March 1980) p. 42.

Sources: For 1980 (1) Wood, Mackenzie & Co., *North Sea Report*, no. 93 (28 January 1981) p. 7. Monthly figures quoted.

(2) Central Statistical Office, *Financial Statisticss*, no. 227. (London: HMSO, March 1981) p. 126. Monthly figures quoted.

(3) Central Statistical Office, *Economic Trends*, no. 329. (London: HMSO, March 1981) p. 42.

price of OPEC was about $34.) In fact, two price series are depicted on this graph. The broken line relates to the old 'posted' (tax reference) price system, described in detail by Park.[23] During the 1970s, however, there was a move away from this system, to one where market prices are not effectively set by the state sales price established by the OPEC countries. Under the former system market prices tended to be below the 'posted' price. Under the latter, periods of relative stability in the price of oil have alternated with periods when substantial increases have ocurred.

Ultimately, the question that is relevant is, 'What price-path of oil during the next two decades should be projected for this study?' Clearly the problem is a long-run one and the changes need to be averaged out so that an overall growth path over twenty years can be obtained. The percentage rate of growth chosen here is an average of 6% rise per year (in real terms) over the last two decades of this century. The actual price path over this period may prove to be below this figure and above it in other years. However, the object of this study is more to demonstrate a methodology in macroeconomic planning for an oil-producing economy. Changes in the price path are simulated in chapter 7 and the impact on the macroeconomic solutions are examined for zero growth and rises of 3% and 8%.

(ii) *Government consumption expenditure*

We perceive government current expenditure as being exogenous in the model and destined to grow after the first four years, 1976 – 9, at a constant rate per annum. Actual values of government consumption are used for the years 1976 – 9 and the exact definition, General Government Final Consumption,[24] which conforms with the official statistics, is applied. Capital expenditure by the government is of course subsumed under the general heading of investment. Values have to be assigned to the exogenous real growth rate of government current expenditure, which increased on average during the 1970s by approximately 2.7% per annum. But the overall figure conceals the fact that the annual rate was slowing down towards the end of the last decade. As a first approximation, a value of 2% per annum is postulated for the real growth rate of government current expenditure after 1980, but the sensitivity of the optimal solution to changes in this value will be further investigated in chapter 7. The entire time path for government current expenditure is shown in Table 5.13

(iii) *The employed labour force*

Our definition and measurement of the total employed labour force was discussed in detail earlier in the chapter when we analysed factor inputs of the domestic non-oil production function.[25] Although the employment figure for 1980 is 24,720, we have projected the size of the employed labour force to stay at around 25,000 during the remaining part of our planning horizon. Hence, no growth is postulated in this variable over the period 1981 – 2000. It is assumed that the year-by-year fluctuations in the level of employed labour around the figure 25,000 will be offset by postulating a constant figure.

The complete time paths for all three exogenous variables specified in the model, i.e., the real sterling price path of oil, the employed labour force and government consumption expenditure, are all shown in Table 5.13.

5.11 The Initial Values of the State Variables

The model embodies three state variables—the domestic non-oil capital stock, K_t, the portfolio of overseas assets, E_t, and the stock of recoverable oil reserves, Q_t, in terms of which the equations of motion are specified in the model.[26]

$$K_t = K_{t-1} + I_t$$
$$E_t = [E_{t-1}/(1+h_t)] + B_t$$
$$Q_t = Q_{t-1} - q_t$$

where I_t = Domestic non-oil investment
B_t = Net investment abroad
q_t = Rate of oil extraction
h_t = Rate of change of dollar/pound exchange rate.

Estimates must thus be assigned to the initial values, K_0, E_0, Q_0, of the state variables.

(i) *Domestic non-oil capital stock*

From the *National Income and Expenditure* book,[27] we obtain a figure of £470.6 thousand million for the gross non-oil domestic capital stock, at 1975 replacement cost, in existence at the end of 1975.

Total gross capital stock	£473,600 million
Gross capital stock in petroleum and natural gas industry	£ 3,000 million
Total gross non-oil capital stock	£470,600 million

This means:

$$K_{1975} = £470,600$$

Table 5.13 The time paths of the exogenous variables: real sterling price of oil, employed labour force and government consumption expecditure

Year	*Real sterling Price of oil (£/barrel)*	*Employed Labour force (Thousand)*	*Government Consumption Expenditure (£ million)*
1976	6.05	24,758	23,581
1977	5.71	24,823	23,334
1978	4.69	24,861	23,865
1979	5.63	25,016	24,334
1980	7.51	25,000	24,820
1981	7.95	25,000	25,317
1982	8.43	25,000	25,823
1983	8.93	25,000	26,339
1984	9.47	25,000	26,868
1985	10.04	25,000	27,404
1986	10.64	25,000	27,952
1987	11.28	25,000	28,511
1988	11.95	25,000	29,081
1989	12.67	25,000	29,663
1990	13.43	25,000	30,256
1991	14.24	25,000	30,861
1992	15.09	25,000	31,478
1993	16.00	25,000	32,108
1994	16.96	25,000	32,750
1995	17.97	25,000	33,405
1996	19.05	25,000	34,073
1997	20.20	25,000	34,750
1998	21.41	25,000	36,450
1999	22.69	25,000	36,159
2000	24.05	25,000	36,882

Notes: Both the real sterling price of oil and government consumption expenditure are expressed in constant 1975 prices.

Sources: Real Sterling Price of Oil 1976 - 1980, see Table 5.2.
1981 - 2000, 6% per annum increase assumed.

Employed Labour Force: 1976 - 1979, *National Income and Expenditure, 1980 Edition.* (London: HMSO, 1980) p. 15.
1980 *National Income and Expenditure, 1981 Edition*, (London HMSO, 1981) p. 15.
1980–2000, constant figure of 25 million assumed.

Government Consumption Expenditure 1976 - 1979, *National Income and Expenditure, 1981 Edition* (London HMSO, 1981) p. 15.

(ii) *Portfolio of overseas assets*
The value quoted for the net stock of overseas assets at the end of 1975 is derived from the figures given in the UK Balance of Payments Statistics, known as the Pink Book,[28] for identified UK external assets and liabilities.

External Assets	£34,751 million
External Liabilities	£36,323 million
Net Stock of Overseas Assets	−£ 1,572 million

As can be seen the UK economy was a net debtor in 1975. Thus

$$E_{1975} = -£1,572$$

(iii) *Stock of recoverable oil reserves*
The stock of North Sea oil is a quantity which is difficult, if not impossible, to define precisely. First there is the problem of an exact definition of 'North Sea'. Does it refer solely to oil finds in the UK territorial waters to the east of Britain or all finds in the UK Continental Shelf? Should oil reserves on the mainland be included as well? Even when an appropriate geographical area has been chosen, there is still the second, and more fundamental problem that the amount of recoverable reserves is not a purely physical notion related to the quantity of oil in place—a quantity which incidentally is not known with any degree of certainty. Rather, the amount of recoverable reserves will depend upon a variety of economic factors and, most particularly, upon the price of oil.

The definition of 'North Sea' adopted here refers to the UK continental shelf. The Department of Energy in their annual publication, known as the Brown Book,[29] report that the regulations under which licenses for petroleum exploration or production are granted apply in respect of territorial waters and areas designated under section 1(7) of the Continental Shelf Act.[30] During 1979 two new areas in the northern North Sea totalling 18,250 square kilometres were designated, bringing the total area of designated Continental Shelf to approximately 643,000 square kilometres.[31] In our model, stock of oil is taken as the total of proven exploitable reserves as recorded by the stockbrokers Wood, Mackenzie and Co[32] (See Table 5.14). This total represents the likely stock of oil that might be recovered from the UK Continental Shelf and comprises production for the following twenty-four fields for which, at the time of writing up the results of this study, development approval had been granted by the Department of Energy.

Of course, more fields are likely to be developed in the future and indeed the Department of Energy has estimated the original quantity of recoverable oil reserves on the United Kingdom Continental Shelf to be in the range of 2,200 – 4,400 million tonnes.[33] Assuming a conversion factor of 1 tonne equals approximately 7.5 barrels,[34] this corresponds to estimates in the range of 16,500 – 33,000 million barrels. The consequences of different assumptions about the ultimate quantity of recoverable oil are examined in chapter 7. Thus in the first instance we use the total figure shown in Table 5.14 i.e.

$$Q_{1975} = 11{,}949 \text{ m/barrels}$$

Table 5.14 Estimate of total recoverable reserves of oil in the North Sea

Field	*Estimate of Recoverable reserves (million barrels)*
Argyll	50
Auk	50
Beatrice	160
Beryl	800
Brae	300
Brent	2,215
Buchan	50
Claymore	404
Cormorant	620
Dunlin	400
Forties	1,845
Fulmar	525
Heather	90
Hutton	250
Hutton North West	280
Magnus	180
Maureen	150
Montrose	90
Murchison UK	318
Ninian	1,200
Piper	618
Statfjord UK	414
Tartan	140
Thistle	500
TOTAL	11,949

Source: Wood, Mackenzie & Co., *North Sea Service: Field Analysis Section*, Bulletins 1979 to August 1980. These bulletins contain estimates for different fields which are continually updated.

5.12 Summary

In this chapter, numerical values have been assigned to the bahavioural parameters of the macroeconomic model specified in chapter 2. These values have been chosen largely on the basis of economic reasoning, although reference has been made, wherever possible, both to available data and to relevant econometric work. The basic parameter set which is used in chapter 6 to generate the optimal solution is summarized in Table 5.15.

It has been stressed throughout this chapter that these estimates are basically first approximations and it is the relative, rather than the absolute, values of these parameters that are important in the context of this study. Simulations around variations in these parameters are carried out in chapter 7. There sensitivity analyses are employed to examine how far the optimal solutions are responsive to alternative values of the coefficients. The analysis in chapter 7 should provide answers to any questions arising about the parameter values assigned in this chapter.

Table 5 . 15 Summary of parameter values

Parameter (1)	*Numerical Value* (2)	*Brief description* (3)
g	112.33	Scale factor in production function
β	0.012	Rate of technical progress
α	0.32	Exponent of capital in production function
γ	0.8	Exponent of labour in production function
ξ	0.95	Proportion of raw material usage to non-oil GDP
θ	0.05	Return on overseas investment
η	0.083	Marginal propensity to consume
ε	0.20	Rate of income-related taxes
σ	0.17	Rate of expenditure tax
Ω_0	0.875	'Non-Oil' exchange rate
Ω_1	0.0075	Scale parameter in exchange rate equation
Ω_2	0.3	Response parameter in exchange rate equation
∇	0.2	Trend rate of growth of non-oil export function
m_0	30,000.0	Scale factor in non-oil export function
m_1	-1.0	Price elasticity of demand for exports
ω	0.5	Proportion of indigenous oil exported
ϕ	0.18	Proportion of raw materials imported
μ	0.10	Proportion of consumer goods imported

6 Numerical Solutions for the Model

6.1 Introduction

This chapter deals with the numerical results obtained from the computer program for our planning problem. The optimal paths for the control variables, which are the non-oil domestic investment and the rate of oil extraction, are determined. These paths are then analysed in order to understand the underlying interplay of the relationships involved.

6.2 Summary of Model Parameters and Other Relevant Preliminary Data

Before outlining the solutions to the model, it is appropriate to list the values of all the parameters and variables measured in chapter 5. The initial values of our eighteen parameters are as follows:[1]

$$g = 112.33 \qquad \xi = 0.95$$
$$\beta = 0.012 \qquad \Omega_0 = 0.875$$
$$\alpha = 0.32 \qquad \Omega_1 = 0.0075$$
$$\gamma = 0.80 \qquad \Omega_2 = 0.30$$
$$\theta = 0.05 \qquad m_0 = 30{,}000$$
$$\eta = 0.83 \qquad m_1 = -1.00$$
$$\omega = 0.5 \qquad \nabla = 0.20$$
$$\phi = 0.18 \qquad \varepsilon = 0.20$$
$$\mu = 0.10 \qquad \sigma = 0.17$$

The starting values for the three state variables for the year 1975 are[2]

$$K_{1975} = £470{,}600\,\text{m}$$
$$E_{1975} = -£1{,}572\,\text{m}$$
$$Q_{1975} = 11{,}949 \text{ barrels}$$

The figure for the initial oil stock is derived from the value of total proven oil reserves only, and does not include potential reserves.

These parameter values and initial stock figures do not comprise all of the model's preliminary data, because there are three exogenous variables whose values are projected separately, and fed into the algorithm.

The exogenous variables in our model are government expenditure, the price of oil and the supply of labour employed in the economy. Government expenditure is set initially equal to actual figures for 1976 – 9, and then a growth rate of 2% has been postulated from 1980 onwards. The path for the price of oil uses international prices (in sterling) for the period 1976 – 80. For the subsequent years it is projected on the basis of an average growth in oil prices of 6% per year.[3] Projecting any constant rate of growth for real oil prices is obviously highly speculative. They have risen substantially, but erratically, in an alternative platform, sharp-rise manner. An assumption of an average 6% rise in oil prices is proposed as a first estimate, to demonstrate the methodology in planning employed in our study. In chapter 7 sensitivity analyses are conducted where alternative price scenarios are visualized. The total labour force in employment is also set at a realistic figure for 1976 – 9, and for the subsequent years is assumed to remain constant. Allowances for the growth in long-term employment will be made in chapter 7. For convenience we have repeated the information in Table 5.13 here (Table 6.1) as a complete summary of the relevant data that are fed into the model before the numerical solutions can be obtained.

6.3 The Solution Procedure

In addition to the inter-temporal state constraints and the terminal constraint, which have been outlined so far, other constraints have also been imposed on the model. These represent the technological constraints, and partly reflect the conditions imposed by the oil companies on the government. Essentially, the companies engaged in oil production in the North Sea apply a very high rate of discount for their investment in this region, as much as 17% – 20% in real terms. This implies that they need to recover their investments as quickly as possible, which in turn means extracting as much oil as is technologically feasible. Most of the companies engaged in exploration in the North Sea in the late 1960s and early 1970s had no restrictions imposed on their rate of extraction. These were companies whose fields came into operation after 1975 and who

Table 6.1 The time paths for the exogenous variables of the model

Year	Government Expenditure (£m) G_t	Real Price of oil (£/barrel) π_t	Labour Force (m) L_t
1976	23,581	6.05	24.8
1977	23,334	5.71	24.8
1978	23,865	4.69	24.9
1979	24,334	5.63	25.0
1980	24,820	7.51	25.0
1981	25,317	7.95	25.0
1982	25,823	8.43	25.0
1983	26,339	8.93	25.0
1984	26,868	9.47	25.0
1985	27,404	10.04	25.0
1986	27,952	10.64	25.0
1987	28,511	11.28	25.0
1988	29,081	11.95	25.0
1989	29,663	12.67	25.0
1990	30,256	13.43	25.0
1991	30,861	14.24	25.0
1992	31,478	15.09	25.0
1993	32,108	16.00	25.0
1994	32,750	16.96	25.0
1995	33,405	17.97	25.0
1996	34,703	19.05	25.0
1997	34,754	20.20	25.0
1998	35,450	21.41	25.0
1999	36,159	22.69	25.0
2000	36,882	24.05	25.0

Notes: See Table 5.13 for the sources and section 5.10 for the assumptions on which these time paths are based.

aimed to recover their investments by the beginning of the 1980s. The real consequence of this clause in the agreement of the oil corporations was that the government had no control over the rate of depletion once a given plot in the North Sea bed was leased out to an oil company.[4]

As far as this study is concerned, the situation described above means that there has to be a minimum constraint on the uplift of oil in the latter half of 1970s. Thus, over the period 1976 – 9 we imposed a minimum constraint equivalent to the value of oil extracted in those years. The figures are as follows:

Year	Actual Values of Oil Extracted in the North Sea (million barrels)
1976	87.0
1977	279.0
1978	395.2
1979	572.3

(The sources for these figures are shown at the foot of Table 6.4)

As it turns out, these four constraints become binding, or nearly binding, in that the values of optimal oil extraction for the years 1976–9 lie very close to the figures shown.

Finally, it should be pointed out that in the macromodel, the functional form of government taxation revenue from the North Sea oil[5] is shown by means of a general function, where l_t^1. To compute numerical solutions to the model, we needed to stipulate an explicit form for this function. In chapter 4 we gave details of the North Sea tax regime, from which it was very clear that the UK oil sector has an extremely complex tax structure. However, if all the various forms of oil taxation were added up–royalties, corporation tax, petroleum revenue tax and supplementary petroleum duty–government revenue would amount to roughly seventy percent of the total oil revenue.[6] This average figure is used for our parameter of the oil-related tax. It can be seen from Figure 4.2 of chapter 4, where a projected rate of taxation over the 1980s and 1990s amounts to roughly seventy per cent. We can now explain the procedure for reaching the optimal solutions. The initial starting values for the algorithm are presented in Tables 6.2a and 6.2b. They show an initial starting point, with a constant rate of domestic investment of £100,000 million and a variable rate of oil extraction. The details of the algorithm used for computing the numerical solutions to our problem are explained in detail in the appendix to this chapter.

The computation algorithm we have used is known as Zoutendijk's Method of Feasible Direction. As explained in the Appendix to this chapter, the algorithm proceeds with computing a solution once it has found a feasible starting point. Since the control variables lie within the feasible domain, the algorithm proceeds to calculate the optimal solutions to the problem. That is to say, the computer program maximizes the global non-oil capital stock, both domestic and overseas, as outlined in chapter 2. The exact optimal point will never be reached using this method, but the algorithm will converge to the optimal point within any desired degree of accuracy. A maximum iteration limit of 150 was chosen on the grounds that good convergence should be achieved well within

Table 6.2a The starting values of the model state variables

Year	*Domestic Non-oil Capital Stock (£ m)* K_t	*Overseas Assets (£ m)* E_t	*Reserves of Oil m barrels* Q_t
1975	470,600	-1,572	11,949
1976	570,600	-81,528	11,859
1977	670,600	-164,323	11,559
1978	770,600	-255,060	11,159
1979	870,600	-348,670	10,559
1980	970,600	-445,394	9,759
1981	1,070,600	-553,698	8,859
1982	1,170,600	-669,719	7,859
1983	1,270,600	-799,616	6,909
1984	1,370,600	-940,177	5,959
1985	1,470,600	-1,096,079	5,059
1986	1,570,600	-1,283,221	4,459
1987	1,670,600	-1,474,003	3,859
1988	1,770,600	-1,683,660	3,259
1989	1,870,600	-1,925,458	2,759
1990	1,970,600	-2,181,954	2,259
1991	2,070,600	-2,464,564	1,759
1992	2,170,600	-2,795,526	1,359
1993	2,270,600	-3,169,651	1,059
1994	2,370,600	-3,597,072	859
1995	2,470,600	-4,061,940	709
1996	2,570,600	-4,549,584	559
1997	2,670,600	-5,135,851	459
1998	2,770,600	-5,740,449	359
1999	2,870,600	-6,498,831	309
2000	2,970,600	-7,253,489	259

Value of Objective Function = −£4,282,888 m

Value of Terminal Constraint = 259.00

the maximum. Figure 6.1 illustrates how the value of the objective function changes as the iterations proceed. It also shows that the algorithm quickly converges towards the solution. Most of the work is done during the first thirty iterations; only very little improvement occurs beyond this point.

6.4 The Numerical Solutions To The Model

The solutions to the model that were reached after 150 iterations are outlined in Tables 6.3a to 6.3e.* Table 6.3a gives the optimal time paths of the three state variables, domestic non-oil capital stock, portfolio of assets overseas, and reserves of oil. Table 6.3b shows the optimal time paths of the two control variables, domestic non-oil investment and the rate of oil extraction, with the latter

Table 6.2b The starting values of the model control variables

Year	*Domestic Non-oil Investment (£ m)* I_t	*Rate of Oil Extraction (m b/year)* q_t	*Rate of Oil Extraction (m b/day)*
1976	100,000	90.0	.246
1977	100,000	300.0	.821
1978	100,000	400.0	1.095
1979	100,000	600.0	1.643
1980	100,000	800.0	2.190
1981	100,000	900.0	2.464
1982	100,000	1,000.0	2.738
1983	100,000	950.0	2.601
1984	100,000	950.0	2.601
1985	100,000	900.0	2.464
1986	100,000	600.0	1.643
1987	100,000	600.0	1.643
1988	100,000	600.0	1.643
1989	100,000	500.0	1.369
1990	100,000	500.0	1.369
1991	100,000	500.0	1.369
1992	100,000	400.0	1.095
1993	100,000	300.0	.821
1994	100,000	200.0	.548
1995	100,000	150.0	.411
1996	100,000	150.0	.411
1997	100,000	100.0	.274
1998	100,000	100.0	.274
1999	100,000	50.0	.137
2000	100,000	50.0	.137

*The superscript * attached to the variables in these tables signify the optimal values of the variables.

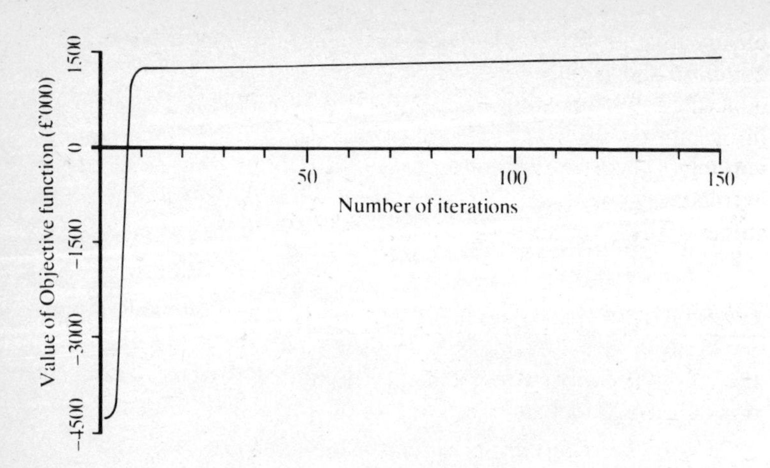

Figure 6.1 Convergence of the objective function as the algorithm approaches a solution

expressed in terms of both million barrels per year and million barrels per day.

Table 6.3c contains the optimal trajectories of some of the endogenous variables, namely, gross domestic product, oil revenue, private consumption and the exchange rate. The time paths of the endogenous variables are determined in relation to the optimal trajectories of the control and state variables; once the model is solved for the time path of the state variables, the time path of the control variables are then determined given the structure and the parameters of the macroeconomic system. Table 6.3d contains time paths of some further endogenous variables. These are income-related tax, expenditure tax, oil taxes, oil exports and non-oil exports.

Finally, Table 6.3e includes the optimal values of non-oil domestic investment, overseas investment, the rate of oil extraction and the values of the state constraint for each of twenty-five years listed. This table repeats the trajectories of the control variables, domestic investment and oil depletion, that appeared in Table 6.3b. This is done in order to compare them with the time path of the balance of payments constraint which is analysed in the following text.

Before examining the optimal time paths as a whole, it is useful to compare the early observations with the actual statistics for these variables. This is done so that (a) the model itself can be checked to ensure it is not generating wildly unreasonable figures; and (b) to compare the actual values of our control variables[7] with those generated by our optimization model. Table 6.4 contains this information, wherein the optimal time paths of the control variables, non-oil domestic investment and the rate of oil extraction, generated from the model are compared with the actual values of these variables in the UK over the period 1976 – 80. Table

Table 6.3a Model solutions for the trajectories of the state variables

Year	*Domestic Non-oil Capital Stock (£ m)* K_t	*Overseas Assets (£ m)* E_t	*Reserves of Oil m barrels* Q_t
1975	470,600	-1,572	11,949
1976	488,806	249	11,862
1977	511,178	122	11,583
1978	534,374	167	11,187
1979	558,201	1,735	10,615
1980	584,759	2,813	10,037
1981	610,836	5,620	9,359
1982	639,470	7,701	8,659
1983	677,412	1,963	7,931
1984	701,234	11,702	7,203
1985	726,054	21,295	6,524
1986	751,604	30,964	5,898
1987	777,566	42,074	5,214
1988	806,450	54,790	4,233
1989	855,142	42,512	4,133
1990	903,834	27,721	4,033
1991	948,241	22,437	3,560
1992	992,648	19,901	3,071
1993	1,037,055	17,465	2,682
1994	1,081,462	15,999	2,315
1995	1,125,868	16,291	1,928
1996	1,177,684	10,736	1,540
1997	1,229,500	5,807	1,203
1998	1,282,296	1,402	866
1999	1,334,149	184	500
2000	1,377,444	12,451	0

Value of Objective Function = £1,389,895 m

Value of Terminal Constraint = 0

Note: Solutions reached after 150 iterations of the algorithm when the system has converged.

6.5 also shows the figures for the optimal and actual values of gross national product and for consumer expenditure. Although gross national product and consumer expenditure are essentially endogenous variables in the model, a comparison of the actual and optimal values of these variables provides a further check on the performance of the model.

The figures for the optimal domestic investment, though higher than the actual values of investment during 1976 – 80, are, nevertheless, of the correct order of magnitude. The optimal rate of oil extraction during this period lies very close to the actual values, reflecting the fact that the minimum constraint imposed for fast

Table 6.3b Model solutions for the trajectories of the control variables

Year	Domestic Non-oil Investment (£ m) I_t^*	Rate of Oil Extraction (m b/year) q_t^*	Rate of Oil Extraction (m b/d)
1976	18,206	87.2	.23
1977	22,372	279.2	.76
1978	23,195	395.3	1.08
1979	23,827	572.4	1.56
1980	26,558	578.1	1.58
1981	26,078	678.1	1.85
1982	28,634	699.5	1.91
1983	37,942	728.1	1.99
1984	23,822	728.1	1.99
1985	24,820	678.6	1.85
1986	25,549	626.3	1.71
1987	25,963	684.6	1.87
1988	28,884	980.3	2.68
1989	48,692	100.1	.27
1990	48,692	100.2	.27
1991	44,407	472.8	1.29
1992	44,407	488.8	1.33
1993	44,407	388.8	1.06
1994	44,407	367.4	1.00
1995	44,407	387.3	1.06
1996	51,816	387.3	1.06
1997	51,816	337.3	.92
1998	52,797	337.3	.92
1999	51,852	365.9	1.00
2000	43,296	499.9	1.36

Note: Solutions reached after 150 iterations of the algorithm when the system has converged.

recovery of investment by the oil companies is binding.[8] Similarly, the optimal values of gross domestic product generated by the model are very close to the actual GDP values. These comparisons of the actual values of the variables with the optimal values determined show that the model is simulating the UK economy reasonably well. It is important to note that in this type of dynamic model—when the system is solved simultaneously for the

Table 6.3c Model solutions for the trajectories of gross domestic non-oil product, oil revenue, private consumption and the exchange rate

Year	*Gross Domestic Non-oil Product (at factor cost)* (£m) D_t^* (1)	*Oil Revenue* (£m) $\pi_t q_t^*$ (2)	*Private Consumption* (£m) C_t^* (3)	*Effective Exchange Rate index* (1975=1.0) S_t^* (4)
1976	96,179	528	63,602	.924
1977	97,988	1,594	65,072	.944
1978	100,805	1,854	66,939	.946
1979	103,601	3,222	68,796	.960
1980	107,135	4,341	71,195	.968
1981	108,640	5,391	72,230	.974
1982	112,571	5,897	74,934	.976
1983	115,604	6,502	77,017	.980
1984	119,169	6,895	79,194	.982
1985	121,941	6,813	81,357	.981
1986	124,785	6,664	83,564	.980
1987	127,688	7,723	85,813	.985
1988	130,632	11,715	88,137	1.000
1989	133,752	1,269	90,630	.939
1990	137,920	1,345	92,990	.940
1991	142,070	6,733	95,255	.980
1992	145,999	7,377	97,688	.984
1993	149,931	6,222	100,215	.978
1994	153,870	6,231	102,749	.978
1995	157,820	6,960	105,323	.982
1996	161,783	7,378	107,965	.984
1997	166,099	6,814	110,646	.981
1998	170,425	7,222	113,355	.983
1999	174,806	8,301	116,118	.988
2000	179,162	12,022	118,970	1.000

Note: Solutions reached after 150 iterations of the algorithm when the system has converged.

twenty-five years, and in the absence of any constraints—the control variables can take any positive values. For instance, since non-oil domestic investment is not restricted in any way, it can take any value between zero and infinity. Thus, for such a close approximation to be achieved, the model must be simulating the structure of the economy fairly well. The contents of Tables 6.4 and 6.5 will be discussed further.

Table 6.3d Model solutions for the trajectories of income-related tax, expenditure tax, oil taxes, exports of oil and non-oil exports.

Year	Income-Related Taxes (£m) l_t^{111}	Expenditure Taxes (£m) l_t^{11}	Oil Taxes (£m) l_t^{1}	Exports Of oil (£m) $\omega\pi_t q_t^*$	Non-oil Exports (£m) V_t^*
1976	19,157	10,812	369	264	34,286
1977	19,600	11,062	1,116	797	33,110
1978	20,162	11,380	1,298	927	33,080
1979	20,722	11,695	2,256	1,611	33,629
1980	21,444	12,103	3,039	2,171	33,839
1981	21,756	12,279	3,773	2,695	34,234
1982	22,570	12,739	4,128	2,949	34,695
1983	23,198	13,093	4,551	3,251	35,292
1984	23,854	13,463	4,826	3,447	35,887
1985	24,505	13,831	4,769	3,407	36,535
1986	25,170	14,206	4,665	3,332	37,280
1987	25,847	14,588	5,406	3,861	38,053
1988	26,547	14,983	8,200	5,857	38,627
1989	27,298	15,407	888	634	38,822
1990	28,009	15,808	942	673	42,157
1991	28,691	16,193	4,713	3,367	42,948
1992	29,424	16,607	5,164	3,688	42,000
1993	30,185	17,037	4,355	3,111	42,712
1994	30,949	17,467	4,361	3,115	43,807
1995	31,724	17,905	4,872	3,480	44,681
1996	32,520	18,354	5,165	3,689	45,413
1997	33,327	18,810	4,770	3,407	46,233
1998	34,143	19,270	5,055	3,611	47,280
1999	34,975	19,740	5,811	4,151	48,134
2000	35,834	20,225	8,415	6,011	48,868

Note: Solutions reached after 150 iterations of the algorithm when the system has converged.

Table 6.3e Model solutions for the trajectories of domestic non-oil investment, overseas investment, rate of oil extraction and values of the state constraint in each time period

Year	Non-oil Domestic Investment (£m) I_t^* (1)	Overseas Investment (£m) B_t^* (2)	Rate of Oil Extraction (m b/year) $-q_t^*$ (3)	Values of State constraints
1976	18,206	1,737	87.2	10,006
1977	22,372	122	279.2	-10,766
1978	23,195	46	395.3	-10,030
1979	23,827	1,570	572.4	-9,075
1980	26,558	1,092	578.1	-9,478
1981	26,078	2,825	678.1	-8,303
1982	28,634	2,097	699.5	-8,804
1983	37,942	-5,715	728.1	-16,787
1984	23,822	9,743	728.1	-1,294
1985	24,820	9,589	678.6	-1,366
1986	25,549	9,654	626.3	-1,264
1987	25,963	11,260	684.6	-238
1988	28,884	13,332	980.3	-0
1989	48,692	-15,818	100.1	-23,340
1990	48,692	-14,740	100.2	-24,686
1991	44,407	-4,140	472.8	-16,635
1992	44,407	-2,469	488.8	-13,423
1993	44,407	-2,546	388.8	-12,709
1994	44,407	-1,465	367.4	-11,801
1995	44,407	349	387.3	-10,293
1996	51,816	-5,524	387.3	-16,164
1997	51,816	-4,957	337.3	-15,129
1998	52,797	-4,394	337.3	-14,807
1999	51,852	-1,212	365.9	-11,993
2000	43,296	12,270	499.9	-75

Note: Solutions reached after 150 iterations of the algorithm when the system has converged.

† No unit of measurement is specified for the values of the state constraints, since they are basically shadow prices and do not reflect market prices.

Table 6.4 Comparison of the actual and optimal figures for the non-oil domestic investment and the quantity of oil extracted

Year	*Actual Domestic Non-oil Investment (1975 prices) (£m)* (1)	*Optimal Domestic Non-oil Investment (1975 prices) (£m)* (2)	*Actual Rate of Oil Extraction (m b/year)* (3)	*Optimal Rate of Oil Extraction (m b/year)* (4)
1976	18,806	18,206	87.0	87.2
1977	18,462	22,327	279.0	279.2
1978	19,259	23,195	395.2	395.3
1979	19,604	23,827	572.3	572.4
1980	19,532	26,558	590.25	578.1

Sources: Col. (1) The figures for this column are calculated as the difference between total gross domestic fixed capital formation and the amount invested in the petroleum and natural gas sector.
The figures are shown in *The National Income Expenditure: 1981 Edition*, (London: HMSO), pp. 17 and 73 respectively.
(3) *Development of the Oil and Gas Resources of the United Kingdom*, London, HMSO, 1981 p. 30. Figures converted from million tonnes to million barrels applying a factor of approximately 1 tonne = 7.5 barrels.
(2) and (4) computed by the model.

Table 6.5 Comparison of the actual and optimal figures of consumers expenditure and gross national product

Year	*Actual Consumers Expenditure (1975 prices) (£m)* (1)	*Optimal Consumers Expenditure (1975 prices) (£m)* (2)	*Actual Gross National Product (1975 prices) (£m)* (3)	*Optimal Gross National Product (1975 prices) (£m)* (4)
1976	64,815	63,602	109,852	107,126
1977	64,583	65,072	110,189	110,657
1978	68,222	66,939	114,177	114,045
1979	71,409	68,796	115,902	118,527
1980	71,454	71,195	113,771	123,666

Sources: Col. (1) and (3) *National Income and Expenditure: 1981 Edition*, (London, HMSO), p. 17.
(2) and (4) computed by the model.

6.5 Economic Implications of the Numerical Solutions

In the analytical solutions in chapter 2,[9] it was suggested that when the balance of payments constraint becomes binding, the optimal policy would be to invest more overseas, but when the constraint is being removed, to invest more at home and to cutback on oil production. Table 6.3e is constructed to verify this solution. In order to compare the behaviour of the two control variables the optimal values of the non-oil domestic investment, I_t^*, the non-oil overseas investment, B_t^*, and that of the oil extraction rate, q_t^*, are tabulated against the values of the balance of payments constraint (which is the state constraint) in each time period. When the constraint has negative values, this implies that the balance of payments constraint is non-binding, when its value is reduced to zero, this reflects the fact that the constraint has become binding. The notions of 'binding' and 'non-binding' can be explained in economic terms in the following manner: when the constraint is binding, it would be optimal to keep imports to a minimum level, i.e., just those which are classed as essential imports; but when the constraint is non-binding there is no pressure on the level of imports.

Referring to our model, the exact form of the solutions states that when the balance of payments constraint is binding in period t, it would be optimal to invest less at home and more overseas in period t-1,[10] and vice versa when the constraint is removed. Also, the solutions state that when the constraint becomes binding, reducing the production of oil would remove the constraint. We shall consider each of these propositions independently in the light of the numerical solutions shown in Table 6.3e, and discuss their implications step by step.

Firstly, concentrating on non-oil investment both domestically and overseas, it can be readily seen that, when the balance of payments constraint becomes binding in 1988, the optimal solutions show less investment at home and more investment overseas in the previous period,[11] namely, 1987. The theory suggests when the constraint is removed more investment should take place at home and less overseas. The numerical solutions reinforce this argument to the extreme, such that in 1989 more investment takes place at home, with none at all overseas. Indeed, the external sector (see column 2 of Table 6.3e) shows negative investment, which implies repatriation of some of the assets accumulated overseas.

Secondly, on the optimal depletion side, when the balance of payment becomes binding in 1988, we can see that it would be optimal to lower the rate of oil extraction. Thus the level of petroleum production is reduced in 1989 and 1990, which permits the state constraint to become non-binding.

In fact, the way the numerical solutions turn out for this particular macroeconomic structure (with this set of parameters and projected time paths of the exogenous variables), the point in time at which the constraint becomes binding marks the division of the planning period into two distinct sub-planning phases. These phases, which happen to be roughly of equal length, mark a clear strategy of accumulation of assets abroad in the former part of the oil life, and expansion of the domestic sector in the latter years of the resource.

An interesting feature of the solutions is that the 'former' and the 'latter' sub-planning phases coincide with the 'pre-peak' and 'post-peak' years of the optimal oil extraction profile. As can be observed from Table 6.3e, petroleum production peaks in 1988. Since that is the thirteenth year of the twenty-five year planning horizon, it indicates that the production peak occurs roughly at the mid-point of the duration of the plan. Thus, the building up of a portfolio of assets overseas becomes synonymous with the pre-oil-peak phase. On the other hand, a rise in domestic investment combined with repatriation of funds from abroad for the growth of the home sector is concomitant with the 'post-oil-peak' phase of the resource. These results are clearly seen in Figure 6.2.

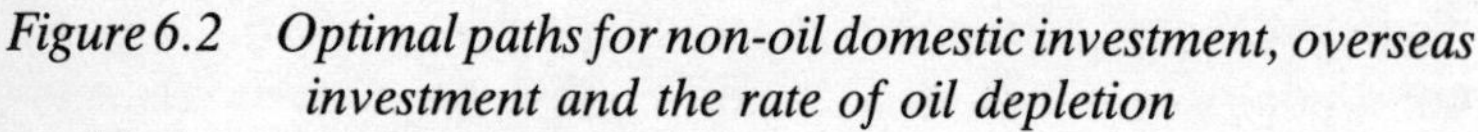

Figure 6.2 Optimal paths for non-oil domestic investment, overseas investment and the rate of oil depletion

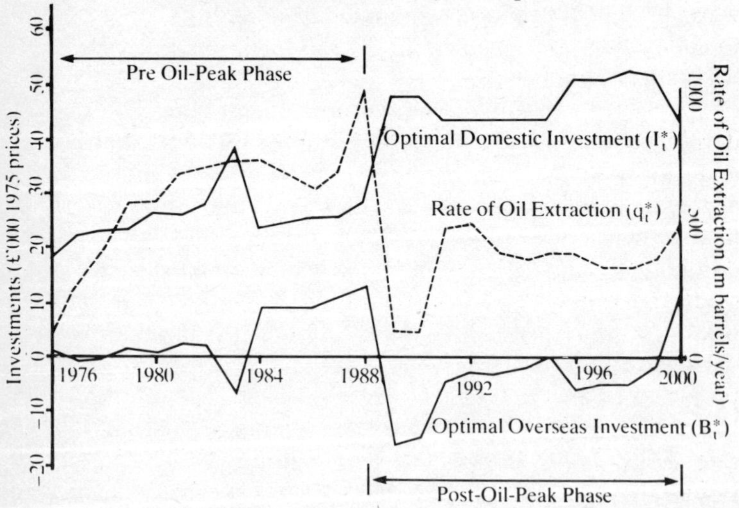

Another important aspect of the pre- and post-oil-peak phases is that the former is associated with a higher rate of depletion than the latter phase, even though the optimal rate of depletion of the pre-oil-peak phase turns out to be lower than the actual rate currently occurring in Britain. This is something which we shall discuss next when we compare the actual UK rate of oil depletion with the optimal path derived in this model. One small point should be clarified here, however. By 'actual' rate of oil depletion, is meant the actual rate of extraction up to 1980/81 (the latest for which figures were available at the date of completion of this book), and thereafter, the planned rate of extraction on the basis of information available from various sources, given the nature of the oil activities in the North Sea.

At this stage it should be pointed out that obtaining information about the future time profile of the UK oil depletion has been an extremely difficult task. The official authorities, namely the Department of Energy, treat the information as confidential. This means that they are unprepared to release any statistics on the planned depletion rate, other than those which they publish in the 'Brown Book'.[12] However, their oil depletion estimates are generally very short-term and have not been going beyond 1984.[13]

Outside the Department of Energy none of the large macroeconomic models of the UK determine any rate of extraction for the oil resource. They all treat it as exogenous. This was discussed at length in chapter 3 when we reviewed all the major models of the United Kingdom.[14] In fact this was one of the reasons why we had to formulate our own macroeconomic model, namely, to endogenize the depletion rate.

Finally, it should be stressed that although there are a few pieces of published research on UK petroleum exploration and development activities, for instance those of Kemp and Crichton,[15] and of Robinson and Rowland,[16] none of them contain an aggregate oil-depletion profile. Their analysis is essentially field-by-field, measuring government taxation revenues on different oilfields. The most important point concerning these studies is that their analyses are primarily done on a microeconomic basis.

However, there have been two independent estimates, giving a projected rate of extraction for North Sea oil on an aggregate basis.[17] These could be regarded as closest to the planned depletion rate for the North Sea oil, particularly in the 1980s. One is produced by Wood Mackenzie[18] and the other by British

Petroleum. The figures of Wood Mackenzie were used in the calculations that appeared in chapter 4. The aggregate production path is derived by summing the projected production profile of all the proven fields in the North Sea.

On the whole, British Petroleum were very reluctant to release any statistics pertaining to their calculations of future production of UK oil. Indeed, they have many scenarios which deal with various uncertainties. However, we did manage to gain access to their projections through a graph they provided, which summarizes their various scenarios. From this graph, entitled UKCS Oil Production and Consumption, 1975 – 2000,[19] we were able to extrapolate their corresponding figures for the aggregate oil production and to derive the values for daily oil production for each year of the period 1975 – 2000. These two projected paths together with the optimal extraction path derived from the model here are presented in Table 6.6 for the purpose of comparison.

As can be readily observed in Table 6.6 the path produced by the model solutions advocates much lower depletion values than those produced by both Wood Mackenzie and British Petroleum for the early and mid-years of the 1980s (up to 1988). In other words, the optimal depletion figures generated by the algorithm here advocate a lower depletion rate as well as a later peak production of oil than those that are likely to be pursued in the UK. The Wood Mackenzie figures show a peak oil production of 2.490 million barrels a day in 1985. The British Petroleum figures indicate a peak production of about 2.4 million around 1984/85. Our model solutions, on the other hand, favour a postponement of the peak until 1988, about four years later.[20] Conversely, our figures indicate a more even path of oil production in the later years—the 1990s—of around one million barrels per day.[21]

Of course, the British Petroleum oil production profile shown in column 2 of Table 6.6 is associated with a scenario that includes only existing fields and those currently under development. BP also has another scenario which includes future fields that are likely to be developed from 1986 onwards. This scenario would not be comparable with our oil production path because the estimates produced by our model are based on 'proven reserves' of 11,949 million barrels, which is roughly comparable with the category of 'existing fields and fields under development' of the British Petroleum company.[22]

Returning to Figure 6.2, the optimal trajectories of oil extraction, non-oil domestic investment and overseas investment look very unsmooth, in fact rather jagged. That is to say, they appear rather volatile, and the question becomes, 'How would the economy as a whole perform under these circumstances?' The best way to answer this question is to inspect the time path of the non-oil GDP, since

Table 6.6 Comparison of the time path of oil production generated by the model with those projected by (1) Wood Mackenzie and (2) British Petroleum (million barrels per day), 1976-2000

Year	*Projected oil Extraction Path produced by Wood Mackenzie* (m/b/d) (1)	*Projected oil Extraction Path produced by BP* (m/b/d) (2)	*Optimal Extraction Path produced by the model* (m/b/d) (3)
1976	.236	.2	.239
1977	.763	.8	.764
1978	1.080	1.1	1.082
1979	1.565	1.6	1.567
1980	1.614	1.7	1.583
1981	1.767	1.9	1.856
1982	2.107	2.1	1.915
1983	2.288	2.3	1.993
1984	2.442	2.4	1.993
1985	2.490	2.4	1.858
1986	2.329	2.3	1.715
1987	2.048	2.0	1.874
1988	1.736	1.7	2.684
1989	1.478	1.4	.274
1990	1.276	1.2	.274
1991	1.085	1.0	1.295
1992	.926	.9	1.338
1993	.813	.7	1.065
1994	.700	.6	1.006
1995	.600	.5	1.060
1996	.526	.4	1.060
1997	.462	.3	.924
1998	.361	.2	.924
1999	.308	.2	1.002
2000	.261	.2	1.369

Sources: Column (1) Wood Mackenzie's *North Sea Oil Service*, various Bulletins 1980/82. More detail is given in chapter 4. These figures are calculated here by summing total output divided by the number of days in a year.

(2) British Petroleum, *UKCS Oil Production and Consumption*, 1975-2000. January 1982. See text for further details.

the object of this exercise is, after all, recycling the oil revenue into non-oil revenue.

Figure 6.3 provides a good exposition of this point. It shows how smooth is the time profile of non-oil gross domestic product, compared with the path of oil extraction rate. (Some would argue, perhaps, that the North Sea oil sector constitutes only a small proportion of gross domestic product.) As further evidence, we compare the trajectories of non-oil investment, both domestic and overseas, with the time paths of non-oil gross domestic product (see Figure 6.4). Again, despite the movements in the paths of I_t^* and B_t^*, non-oil gross domestic product remains stable and fairly smooth.

Figure 6.3 Comparison of the optimal paths of non-oil gross domestic product and the rate of oil extraction

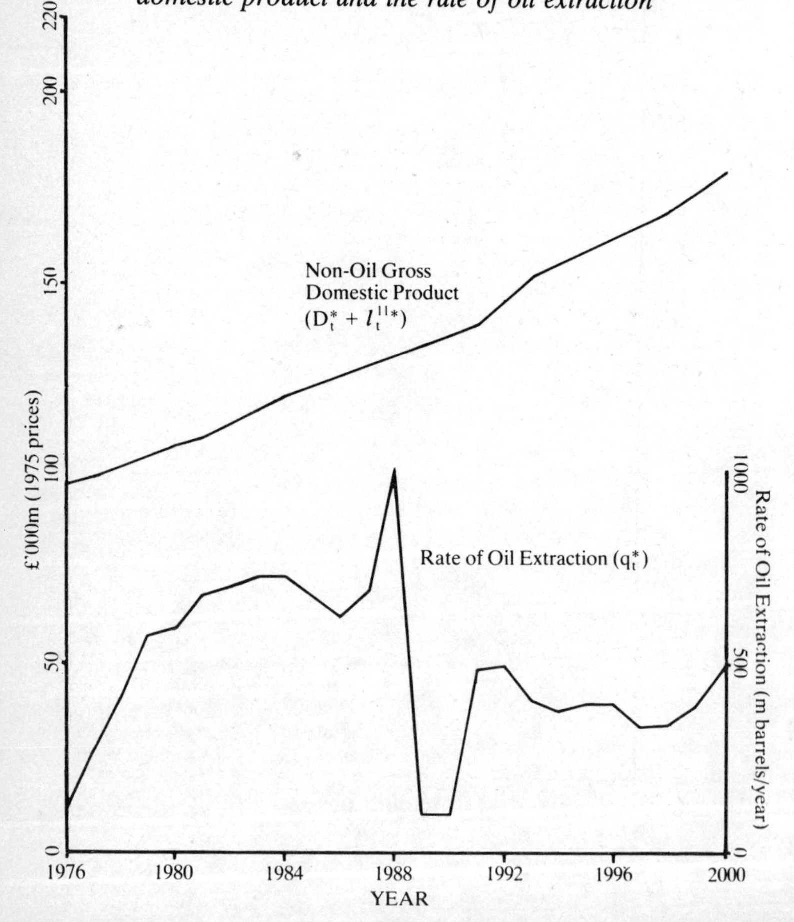

In fact if one were to measure the total impact of the oil revenue on the UK economy, the solutions generated by the model would indicate an overall growth in the economy of 86%. In other words, if optimal policies were pursued, the effect of North Sea oil on gross domestic product would be a permanent increase of some eighty-six per cent. This would be equivalent to raising the rate of growth of gross domestic product by 2.5 per cent per year over the period 1976 – 2000.

Figure 6.4 Comparison of the optimal paths of gross domestic product, domestic investment and overseas investment

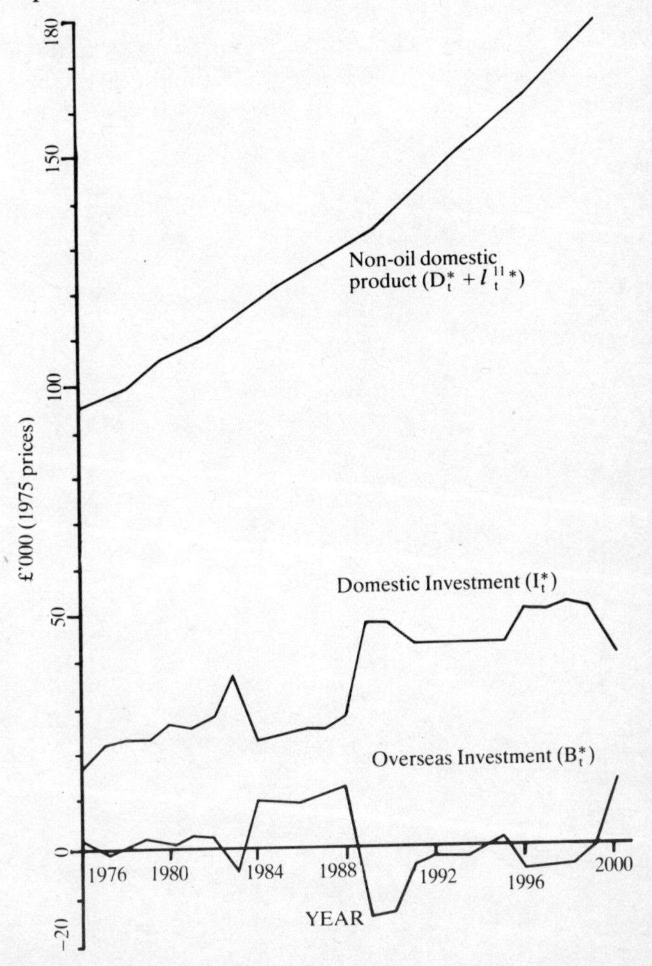

As a further test of the model's performance, we can scrutinize the time paths of our stock variables, domestic fixed capital stock and overseas assets. The time paths of these two variables are shown in Figure 6.5. As can be readily observed, the domestic capital stock is characterized by a smooth upward-sloping curve. The foreign asset function shows a period of building-up reserves overseas, followed by gradual repatriation. Indeed the curves associated with these two variables clearly bring out the pre-oil peak and post-oil peak phases that were discussed earlier.[23]

Domestic capital stock shows a gradual rise up to 1988 coupled with fast accumulation of reserves overseas. The trend reverses after 1988, where rapid accumulation of capital occurs at home and funds are repatriated from abroad. Of course the magnitude of overseas assets is considerably smaller than that of the domestic sector. For this reason we have chosen a different scale on the right hand axis of Figure 6.5 to emphasize the shape of the E_t function.

Figure 6.5 Comparison of the optimal paths of domestic non-oil capital stock and overseas assets

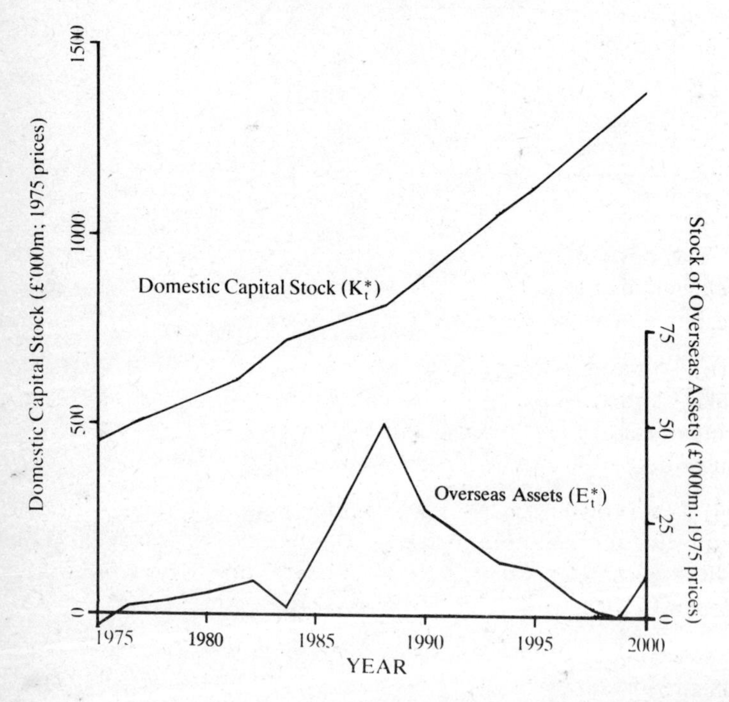

Another important aspect of our planning problem is the development of non-oil exports. The numerical solutions of the model provide us with the optimal trajectories of non-oil exports and exchange rate. This is mapped out in Figure 6.6. As may be recalled from the specification of the model,[24] non-oil exports were shown as a function of the lagged values of the exchange rate, where the latter was, in turn, a function of the size of oil revenue. The optimal time path of this variable is also shown in Figure 6.6.

Figure 6.6 Comparison of the optimal paths of non-oil exports and the exchange rate

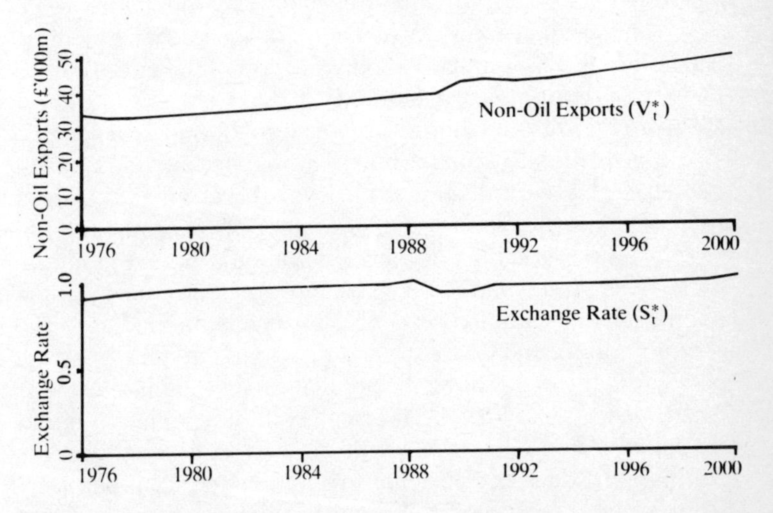

The development of the non-oil export sector is particularly relevant in the light of the non-replenishable nature of the oil resource. This sector would have to fulfil two roles:

(i) To pay for future imports of fuel, whatever form they may take,[25] when North Sea oil is exhausted. This would be roughly equivalent to the amount which is currently saved through import substitution during the life of this resource.

(ii) To provide for the extra foreign exchange gained through exporting a certain proportion of the petroleum produced in the UK, which would disappear when the North Sea reserves are depleted.

Of course, additional fields may be discovered and come on-stream before the turn of the century. But these are all areas of uncertainty and we shall deal with what is certain for the moment, in order to complete the analysis. Any of the assumptions can be relaxed, and their consequence on the overall solutions investigated. This is the subject matter of chapter 7, in which sensitivity analyses are carried out.

As can be seen from Figure 6.6, non-oil exports, V_t^*, grow throughout the 1980s and 1990s. The low depletion rate of oil, as shown in Table 6.3b ensures a fairly steady path of the dollar/sterling exchange rate, hovering around the 1.80 – 1.90 million barrels per day mark. There is of course a hump in the level of non-oil exports, corresponding to a dip in the exchange rate over the period 1988 – 92. This is best explained by looking at Table 6.3d. In 1988 the balance of payments constraint becomes binding, which in turn means a cutback in oil production. The table shows a drop in the rate of oil extraction in 1989 and 1990.

As was seen from equation 2.17,[26] the level of the exchange rate is a function of the size of oil revenue. Thus a fall in oil production in 1989/90 will ensure a reduction in the level of exchange rate, which will then, (through equation 2.16 of chapter 2), boost the level of non-oil exports. The rise in the level of non-oil exports will in turn help to ease the balance of payments pressure, and in the subsequent years the constraints become non-binding. The mechanisms of the model are therefore brought out quite clearly in Figure 6.6. Table 6.3d shows the optimal values of non-oil exports over the years 1976 – 2000. This implies that, if optimal policies were pursued, non-oil exports would grow by 42 per cent over the whole period, which is equivalent to about 1.4 per cent per year.

The numerical solutions for changes in taxation variables, level of private consumption, government expenditure, and so on, will be discussed in chapter 7 in the context of sensitivity analysis. This is because the implications of the solutions can best be brought out when variations in the size of the parameters or projected paths of the exogenous variables (say, the price of oil) are considered, and the effect on the overall solutions investigated. We shall show that the solutions can be sensitive to some alterations in the parameters and projected paths of the exogenous variables, but remain fairly robust with respect to others.

6.6 Conclusions

The central purpose of this chapter has been to show how numerical solutions can be obtained for the model. The results shown here verify the analytical solution of the model presented in chapter 2. The rate of depletion of petroleum oil which emerges in our numerical solutions in the 1980s is lower than that projected by the Department of Energy and some of the oil companies. Moreover, our solutions show that peak petroleum production should occur, optimally, in the late 1980s (around 1988), which is later than that projected by any of the sources cited. Finally, as far as recycling of the oil revenue is concerned it would be optimal to invest more overseas, even at a lower rate of return, during the 'pre-oil-peak' period and more at home in the 'post-oil-peak' period.

Appendix
The Optimization Algorithm

For the purpose of the computation of the numerical solutions, we made use of the Zoutendijk Method of Feasible Directions.* This method was found to be suitable for fixed time problems with inequality constraints on the state and control variables, as well as constraints on the terminal states.

The algorithm is designed to solve problems of the general class that may be written as:

$$\text{Max } \{f^0(z) \mid g^j(z) \leqq 0, j = 1, \ldots m\}$$

where f^0 (.) is the value of the objective function over the whole period; g^j (.) are the constraints, m being their total number; and z is a vector of control variables whose number of elements will depend on the number of control variables and on the length of the time interval representing the control horizon. Without going into too much technical detail, we will now explain those features of the algorithm that throw light upon its working and help to explain the mechanism.

The algorithm computes a solution to the problem in two parts. It first finds a feasible solution that satisfies all the constraints, and then finds an optimal feasible solution. Thus, for the purpose of applying the algorithm, a control strategy must first be introduced as a possible solution to the problem. The program then considers the strategy and, as a result of this examination, two situations can arise:

*See Polak, E. (1972), 'A Survey of Method of Feasible Directions for Solutions of Optimal Control Problems',*IEEE Transactions on Automatic Control*, Vol. AC-17, No. 5, October.

either (a) the strategy satisfies all the constraints, in which case the initial values are said to be feasible;

or (b) some of the constraints are violated, making the control strategy infeasible.

The latter is a condition that happens more frequently, since a feasible solution may not be known initially. At this point, the algorithm's first task is, therefore, to find a feasible solution. If there exists a feasible solution to the objective function, then it can be found by solving:

$$\text{Min } \{z^0 \mid g^j(z) - z^0 \leqq 0, j = 1, \ldots m\} \qquad \text{A6.2}$$

where z^0 is the feasible control vector; g^j (.) still remain as the constraints, and z as the vector of the control variables. Once this sub-problem is solved and a feasible starting point is obtained, the algorithm finds a local maximum of f^0(z) within the feasible solutions. Reaching an optimal solution is quite cumbersome, and involves going through a large number of small steps.

The optimal feasible solution is reached by calculating a search direction which, near to the initial point, lies within the set and along which the value of $f^0(z)$ is guaranteed to maximize. Therefore, the search direction is carried out by first calculating what is called the ε-binding constraints. These are constraints that are binding or within a distance ε of the initial feasible point. Then the search direction is calculated using linear programming, such as the simplex method. Each time the estimate of the vector z is updated, an iteration occurs. Maximization, therefore, takes place by a series of iterations and a maximum is determined through a convergent sequence.

7 Simulations and Sensitivity Analyses

7.1 Introduction

The last chapter outlined the numerical results for the basic model, with the assigned set of parameters from chapter 5. It is, however, useful to ask, 'How might the optimal solutions change if some of the basic parameters were varied?' This question is particularly useful when considering such variables as government expenditure, the income tax rate, the expenditure tax rate or the levels of private consumption. All these variables are to some extent under government control, and changes in them could affect oil depletion policies and investment of the revenue. It is important, therefore, to know precisely what the effect of changing these variables would be.

The question can be answered fairly easily by conducting simulations against the basic model and comparing them with the solutions outlined in the last section. In each simulation we vary only one parameter or relationship at a time. This will allow us to investigate the effect of these changes as well as studying how robust the basic model solutions are to variations in the basic model structure.

It was thought best to avoid tabulating the numerical solutions for each simulation as was done in chapter 6, for that would be a lengthy and somewhat tedious procedure. Rather, it was decided to explain the main economic impact of each sensitivity analysis and highlight its underlying policy implication.

7.2 Depletion Sensitivity to Legal Constraints

The basic solution, outlined in the last chapter, has the addition of four constraints, each imposing a minimum level on the oil extraction rate in the first four years of the life of North Sea oil, 1976 – 9. Those were imposed on the grounds that it was the policy of the oil companies to recover their investments in the North Sea as quickly as possible, and that they were lifting as much crude oil as was technologically feasible. While this approach is undoubtedly correct for the basic solution, it is interesting to discover how oil depletion would have been affected if the government were free to choose the rate of extraction. What this implies is that the rate of petroleum production in the model would have been constrained only by the size of the proven oil reserves. The optimal oil depletion path is then determined on the basis of the macroeconomic considerations, given the price of oil.

The most immediately striking point to be made about this simulation is that the oil depletion path in the first four years changes radically. As one might expect, the optimal rate of oil extraction is lower than the actual. In the first two years, 1976 and 1977, the optimal figures, though positive, lie below those of actual values of oil production. In the next two years the production values associated with the optimal rate become very low and rather negligible. Suffice it to say, that the reason for these lower values of depletion, in particular, during 1978 – 9, lies in the path of oil prices in those years, which actually fall in real terms.[1]

This simulation is a good test of the behaviour of the model, for it shows that the optimal oil depletion path is determined in relation to the price of petroleum. Since the real price of oil falls during the period 1976 – 9 (with the lowest value in 1978) and rises in 1980, it is found that the value of q_t^* is lowest in 1978 and accordingly rises in 1980. Of course the rate of oil extraction is also determined on the basis of macroeconomic considerations as shown in chapter 2. Nevertheless, the role of oil prices in the extraction path is clearly brought out in this exercise.

Overall, the results indicate that if the British economy had had the choice in determining the amount of crude pumped out by the oil companies, it would have been optimal to extract less in 1976 – 9 and to produce only very little oil in 1978 and 1979. But, in practice, those optimal values could not have been attained by the British economy in the years in question. This was because, once a given plot on the North Sea bed had been leased out to an oil company, the government could not exert any further control over the rate of depletion.[2] After the initial periods, the oil depletion path generated in this simulation run turns out to be similar to the basic solution of chapter 6.[3] That is to say, since the same real price path of oil in the long term is postulated, the oil depletion path generated is very similar to that of the basic run.

7.3 Changes in Government Expenditure

Possibly the greatest temptation facing any government when a large new source of revenue is presented to it, is simply to use that revenue to finance a growth in the overall government expenditure. The model has been solved for an exogenous path of government expenditure which starts from the same initial figure and then rises at 3% a year instead of the original figure of 2%.

The effect of this one per cent increase in the growth rate of government expenditure is most interesting. Oil extraction seems to have been brought forward, so that the increased government expenditure can be financed by the additional oil revenue. Peak oil production is now at a slightly lower level, compared with the basic solution in chapter 6 when the growth of government expenditure was set at 2% per annum. Moreover, oil production for the last ten years of the period (1991 – 2000) is consistently below that determined in the basic solution. Similarly, domestic non-oil investment is lower in those same years, except in the last year which marks the end of the period. This causes a lower value of the objective function, which is the maximum value of non-oil wealth accumulated over the life of oil. The reduction in the value of non-oil wealth is also caused by the lower values of overseas investment in the earlier years (up to 1983), which in turn is due to the rise in the level of government expenditure.

7.4 Changes in the Level of Aggregate Consumption

The main reason for undertaking a sensitivity analysis on the average propensity to consume is to see the effect on the complete solutions of a higher level of aggregate consumption in the economy. In the basic solutions (chapter 6), the average propensity to consume was set equal to an average of 83% of disposable income. Economic theory would suggest that the immediate effect of higher consumption would be a lower level of investment. On the dynamic side, a direct increase in consumer expenditure would in turn mean a reduction in the overall level of non-oil wealth over the life of the oil.

This is verified in our numerical solutions. The value of the average propensity to consume here was changed to 90% in order to simulate the effect of a consumption boom resulting from the advent of the substantial oil revenues. As a result of this change the level of domestic non-oil investment becomes substantially lower, with a consequent large reduction in both the terminal value of the domestic capital stock and the overall value of non-oil wealth, which is basically the value of our objective function in the model. Both foreign investment and the rate of oil extraction are affected. The level of the former is on the whole reduced, while the path of the latter is altered so that the oil production rate is stepped up in the pre-oil-peak years and reduced in the post-oil-peak years. Thus, the effect of higher government expenditure on oil depletion is quite similar to a higher level of aggregate consumption: they both use up more resources of the economic system. Higher oil revenue, and hence, a higher depletion rate would be needed to finance the extra expenditure in the earlier years. As far as the later years are concerned the invested oil revenue of the earlier years should generate sufficient returns to finance the extra expenditure.

A further simulation was undertaken for the case where the average propensity to consume was reduced, and once again a priori expectations as discussed earlier were fulfilled. The results turned out to be the converse of those for an increase in consumption. But, for the sake of space, these results will not be discussed further.

7.5 Changes in the Rates of Income-related and Expenditure Taxes

The simulation of the previous section shows the effect of changing the level of consumption, but not how this eventuality might be achieved. In practice, the main, direct instruments by which the government could affect the level of consumption would be the tax rates. Broadly speaking, lowering the rate of either the income related taxes, or expenditure tax[4] would give rise to similar effects as increasing the average propensity to consume. These predictions were verified when examining the results of the simulations showing the effects of a reduction both in the rate of income-related taxes and in expenditure tax. Each of these is discussed in turn.

(i) *Income-related taxes*

To examine the effect of a reduction in this variable, we lowered the rate of income-related taxes from an average of 20% (used in chapter 6) to 17%. The model solutions show clearly that the level of aggregate private consumption is enhanced throughout our planning period compared with that of the 'basic' case shown in Table 6.3b.

Conventional economic theory would suggest that, as consumption rises (due to the fall in income tax), less funds are available for investment expenditure. This is verified in the solutions to this simulation, where the overall size of gross domestic fixed capital formation (GDFCF), I_t^*, declines as the rate of income-related taxes is lowered. Bearing in mind that our definition of I_t^* includes both private and public sector investment, the drop in the level of GDFCF would occur in the public sector,[5] for the less tax revenue is available, the less government funds can be released for capital expenditure. This will in turn result in a reduction in the aggregate level of investment, I_t^*.

The effect of the change in income-related taxes on the level of consumption becomes cumulativc over time. As more resources are allocated to direct consumption and less investment takes place, the growth of output slows down. This implies proportionately less income and less investment as time progresses. What emerges is a gradual closing of the gap between the two consumption paths (i.e., of the solutions here and that of chapter 5) and a widening of the investment paths in these simulations.

(ii) *Expenditure tax*

Rather similar results (in relative terms) are reached for a reduction in the rate of expenditure tax—17% in the basic case, to 13%. Again, compared with the numerical solutions of chapter 6, private consumption is enhanced and non-oil domestic investment is reduced.

The oil revenue path reveals that petroleum depletion is brought forward and the level of production is enhanced during the pre-oil-peak period. The rise in oil revenue in the earlier years is required in this instance to finance the loss in the tax revenue during those years. In the later years the income from the non-oil sector (both domestic and overseas) is assumed to grow and compensate the shortfall in oil revenue, which would be caused by the lower production levels.

On the whole, the time path of oil depletion appears fairly robust, although there seems to be some sensitivity to the changes in the other taxation variables. Indeed, the simulations carried out in this chapter show that the optimal rate of petroleum extraction remains reasonably robust when changes occur in some of the other macroeconomic variables such as government expenditure, aggregate consumption expenditure, etc.

7.6 Changes in the Parameters of the Exchange Rate and Export Functions

Oil revenues can affect the level of non-oil exports through changes in the exchange rate. The long-run relationships between these two variables are specified through the pair of equations 2.16 and 2.17 of the macromodel in chapter 2.[6] When quantifying the parameters of these equations in chapter 5, we modified equation 2.16 to allow for the trend rate of growth in non-oil exports.[7] Since the addition of this time trend does not affect the analytical solutions of chapter 2, it is not included in equation 2.16 there. However, it is important to include it in this equation for arriving at correct numerical solutions.

The pair of equations which capture the effect of North Sea oil on the exchange rate are reproduced here to remind the reader of their functional form.

$$S_t = \Omega_0 + \Omega_1 (\pi_t q_t{}^{\Omega})_2$$
$$V_t = m_0 (1 + \nabla)^t S_{t-1}{}^{m}{}_1$$

The strength of these relationships obviously depends on the values of the six parameters which appear in them. The magnitude of any one, or combination, of the parameters may be varied during a sensitivity analysis. The permutations are, therefore, numerous. Here, we examine each equation separately, and hence, variations in their respective parameters.

(i) *The exchange rate function*

This function depicts the long-run effect of oil revenue on the level of effective exchange rate. This means that both the price of oil and the quantity of oil extracted can influence the effective exchange rate. First it was decided to reduce the value of Ω_1, which is basically a scale parameter, from 0.0075 (used in chapter 6 for the basic solutions) to 0.005, while the values of all the other parameters are held constant. The effect of the change in the exchange rate function is shown in Figure 7.1. The variable considered is, in fact, the effective exchange rate, similar to that shown in column 4 of Table 6.3c of chapter 6. Clearly, as a result of this alteration, the effect of oil revenues on the exchange rate is weakened since the value Ω_1 is lowered. The question that arises, then, is; what are the consequences of a weaker response of the exchange rate to the level of North Sea oil revenue? The key variables to be considered here are non-oil domestic investment, overseas assets and the rate of oil extraction.

Comparing the rate of oil extraction with that associated with the higher value of Ω_1 in chapter 6, it is found that the values of q_t^* in the 1980s here, are reduced. Conversely the values of q_t^* in the 1990s are enhanced. The alteration in the path of the oil extraction is due to the second-rounds of effect which the exchange rate has on non-oil exports. In other words, while the oil extraction and, hence,

Figure 7.1 Changes in exchange rate function when Ω_1 is reduced

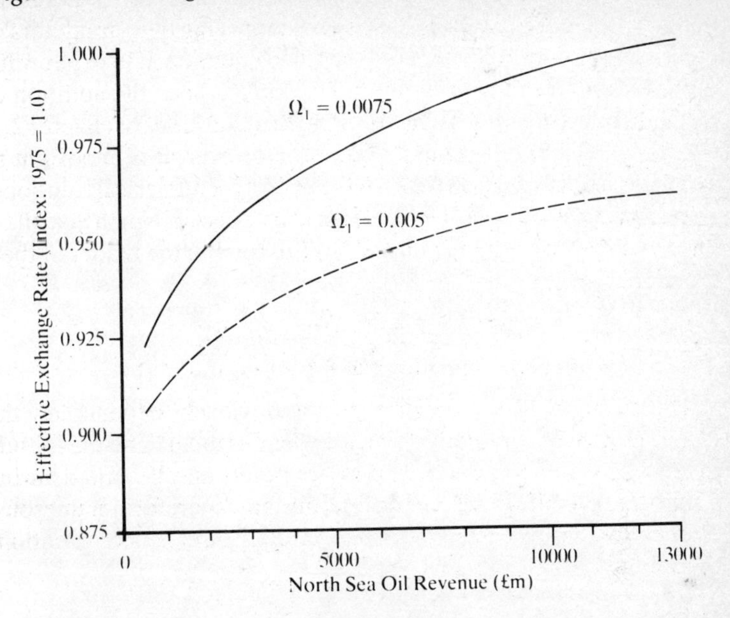

the oil revenue affects the exchange rate, this in turn influences the level of non-petroleum exports. However, as the influence of the oil revenue on the exchange rate deteriorates, non-oil exports do not suffer abroad by as much and remain competitive. This implies that oil extraction does not have to rise by as much in the earlier years[8] to compensate for the adverse effects on non-oil exports. Higher production can be left to the later years when the price of oil is higher. The deferment of oil production will help create a higher overall non-oil wealth over the life of oil, and this is reflected in the value of the objective function.

The level of overseas assets E_t^*, is another variable whose level is directly affected by the exchange rate through equation 2.9 of the macromodel.[9] This is reproduced here to clarify the nature of its effect.

$$E_t = B_t + [E_{t-1}/(1+h_t)]$$

$$\text{where } h_t \equiv \frac{S_t - S_{t-1}}{S_{t-1}}$$

Clearly, h_t represents the percentage change in the exchange rate. Thus, as the level of the effective exchange rate rises, the value of overseas assets in sterling terms falls, and vice versa when assets rise. In this simulation, when oil revenue is postulated to have a weaker impact on the exchange rate it implies that the value of overseas assets in sterling terms is not eroded by as much. Hence, the size of the portfolio of assets abroad is larger compared with that of chapter 6, where $\Omega_1 = 0.0075$.

We have also considered increasing the value of Ω_1 to 0.01, which would be equivalent to strengthening the effect of oil revenue on the exchange rate. The results are more or less symmetrical to the ones presented above. Another simulation run was carried out which also has the effect of strengthening the impact of oil revenue on the exchange rate. This simulation considered the sensitivity of another parameter of the exchange rate equation, namely Ω_2. Here, we increased the value of Ω_2 from its original value,[10] of 0.30, to 0.32. In this way, it was possible to see the consequences of a stronger response of the exchange rate to the size of North Sea oil revenue in the UK. As in the previous case the key variables considered are non-oil domestic investment, overseas assets and the rate of oil extraction. From Figure 7.2, we can see that the impact on oil extraction is the opposite of the previous case, where the response of the exchange rate to oil revenue was weak.

Compared to the base case, where $\Omega_2 = 0.30$ (see Table 6.3c of chapter 6), oil production is brought forward in time. The rate of extraction is much higher in the pre-oil-peak period and much lower in the post-oil peak phase. The rationale for this result is the same as that explained before, namely the secondary effect of the oil revenue on non-oil exports via the exchange rate. When the exchange rate responds strongly to the level of petroleum revenue, more oil needs to be extracted in the earlier years, in part to be exported. This would be necessary to compensate for the loss of non-oil exports that would be caused by the rise in the exchange rate.

The lower oil production in the post-oil-peak phase turns out to be optimal when the response of the exchange rate to oil revenue is strong. There are two reasons for this. First, since the price of oil is postulated to rise over this planning period, the higher prices would compensate for the drop in production, and thus oil revenue would not fall by as much. Second, the lower oil production would lead to a lower exchange rate and non-oil exports would then rise. Of course, the overall size of the objective function, which is the accumulation of non-oil wealth, is reduced.

Figure 7.2 Changes in exchange rate function when Ω_2 is increased

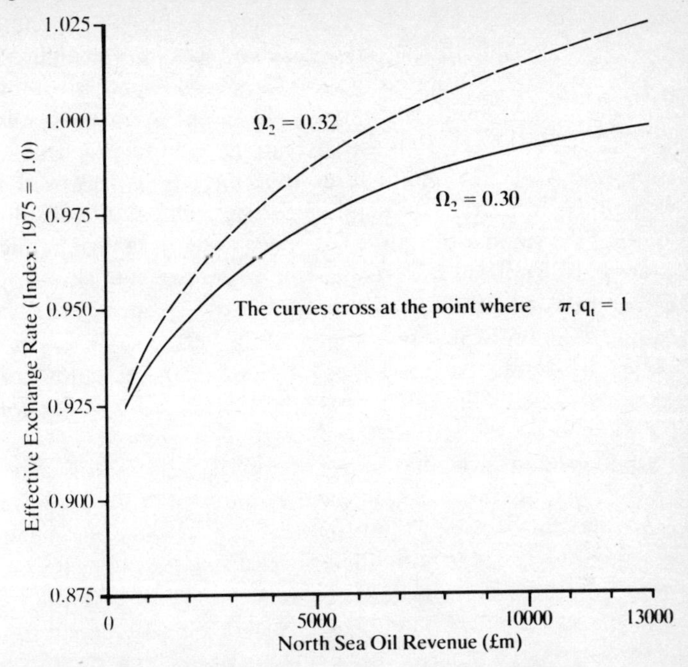

The question becomes then why should one not reduce the rate of petroleum production in the earlier years when the response of the exchange rate to oil revenue is strong? The answer lies in the smooth growth path of exports. Here, the paths of non-oil exports in our numerical solutions, when $\Omega_2 = 0.30$ and when $\Omega_2 = 0.32$—the case of a weaker and stronger response of exchange rate to oil—were compared. Evidently, this policy ensures that non-oil exports grow on a smooth path and reach the same size by the end of the planning horizon as in the base case despite the higher level of the exchange rate. Under these circumstances, if higher oil production occurred in the later years the rise in exchange rate would have adverse effects on a larger volume of non-oil exports than if oil production were higher in the earlier years, and thereby, the overall size of the non-oil wealth would be reduced. Another simulation run was made for the case where the value of Ω_2 was reduced from 0.30 to 0.28, and the results were again the opposite of the above case when Ω_2 was increased to 0.32.

(ii) *The non-oil export function*

This function, as already shown, depicts the relationship between the effective exchange rate and the level of non-oil exports. The function portrays an inverse relationship, whereby non-oil exports are a lagged function of the effective exchange rate. We conducted sensitivity analyses on the parameter m_1 and m_0 and ∇ unchanged. Originally, we had postulated $m_1 = -1.0$, as in chapter 5.[11] Here, we have performed simulation runs for the values of $m_1 = -0.9$ and $m_1 = -1.1$. The effects of these alterations on the shape of the function excluding the element of the time trend ∇ would be equivalent to those shown in Figure 7.3. The greater[12] the absolute value of m_1 the steeper the curve becomes, and this in turn implies, the more elastic non-oil exports become with respect to changes in exchange rate. Bear in mind that Figure 7.3 is essentially a Walrasian diagram, where quantity is on the vertical axis and price on the horizontal, and it should not be confused with the conventional Marshallian diagrams where price is on the vertical axis and quantity on the horizontal axis.

The values of different elasticities of non-oil exports to exchange rate were calculated in our numerical solutions. However, we can demonstrate this analytically in Figure 7.3. The curve associated with the $m_1 = -0.9$ is the more inelastic case and the one associated with the $m_1 = -1.1$ is the more elastic case. The more inelastic case reflects the notion that, when the exchange rate appreciates, export demand does not fall significantly. Conversely, the more elastic case portrays that a gain in exchange rate would lead to a greater fall in exports.

Accordingly, our numerical solutions show that the more elastic export function culminates in a higher value of non-oil exports, V_t, by the end of the planning period than that associated with the less elastic function. On the other hand, when non-oil exports are less elastic the value of accumulated non-oil wealth (shown as the value of our objective function) is greater than in the more elastic case.

The rationale for these results can be best explained by recourse to the path of some of the other variables of the model solutions. Here we examined the optimal depletion policy associated with each elasticity value. When the non-oil export function is less elastic (when $m_1 = 0.9$), a lower extraction rate in the pre-oil-peak period is recommended and a higher rate in the post-oil-peak phase. Note we are comparing these solutions with those of the base case shown in Table 6.3b of chapter 6. Alternatively, when the demand for oil is more elastic a higher extraction rate is viable in the earlier years and a lower one in the later years.

The reasoning behind these rates of petroleum production lie in their impact on the effective exchange rate. When non-oil exports

Figure 7.3 *Non-oil exports as a function of effective exchange rate with different values of elasticity*

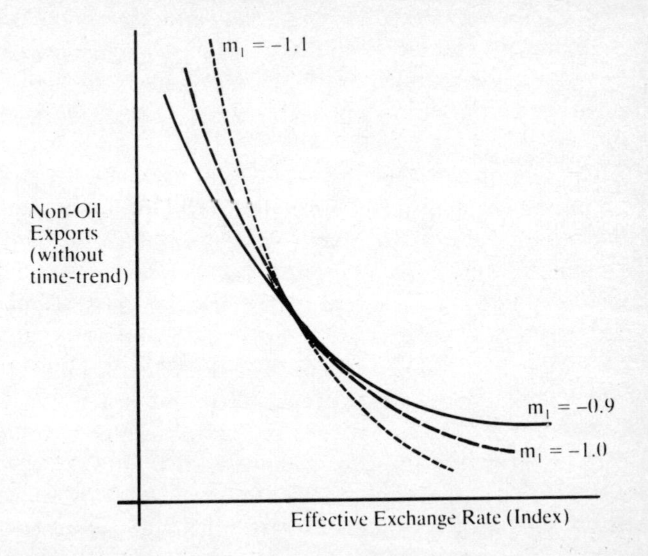

are fairly inelastic to the exchange-rate movements, this implies that demand for exports does not fall when exchange rate rises. So additional oil production for exports would not be required in the early years to compensate for the loss of non-oil export revenue. So higher production than the base case would occur in the later years when oil prices are higher. On the other hand, as non-oil exports are likely to suffer when overseas demand for UK goods and services is fairly elastic, oil production would have to be stepped up in the earlier years for export purposes, to bridge the gap caused by the decline in non-oil export revenue.

Suffice it to say that the optimal policies proposed here are in pursuit of maximum non-oil wealth over the life of oil. Given this objective, less wealth can be accumulated over the life of oil when the demand for non-oil exports is more elastic.

7.7 Changes in the Projected Price Path of Oil

In section 7.2, it was suggested that one of the important determinants of the rate of oil extraction was the exogenous path of oil prices. The lower the rate of oil-price increase in the future, the more attractive will be the option of immediate extraction of oil, and investment of the revenues so derived, in some productive activity. Similarly, if a high rate of price increase is anticipated, the objective of maximization of global wealth will be best served by depleting the oil stock more slowly. These assertions will now be tested by simulating the model for three scenarios; when no price increase has been assumed as well as 3% and 8% annual increases in the real sterling price of oil after 1980. Indeed, the oil depletion path is affected quite considerably in each run and they differ from one another.

In the simulation run where a 3% per annum increase is assumed, the optimal oil extraction level is higher in the pre-oil-peak years compared to the base case of chapter 6. Lower values of extraction are optimal for the later years. Average annual production is around 200 million barrels a year in the 1990s. In contrast, the 8% path yields a much lower extraction rate in the earlier years, starting with 1.7 million barrels in 1976. The peak does not occur until 1992 when the optimal rate of extraction is 770 million barrels. What emerges from these two scenarios is an upward trend in production with the 8% growth path, and a downward trend with the 3% growth rate.

As for the paths of petroleum extraction when there is no rise in the real price of oil over our planning horizon, the response found in the optimal extraction was quite considerable. Oil production is brought forward into the earlier years, according to the optimal solutions, and the resource is virtually exhausted by the early 1990s. It should be borne in mind that all these assumptions about the growth of the price of oil run to the end of the century.

7.8 Changes in the Rate of Technical Progress

One of the factors influencing the production of domestic output is the rate of technical progress. This parameter, which appears as β in our production function[12] and is reproduced below,

$$D_t = g(1+\beta)^t K_{t-1}{}^{\alpha} L_{t-1}{}^{\gamma}$$

will affect the rate of return on domestic investment. Hence, changes in the parameter will in turn influence the allocation of resources between overseas and domestic investments. Clearly, the greater the rate of technical progress, the greater the marginal product of the factor input. Economic theory would suggest that an increase in the rate of technical progress at home would call for larger domestic investment. This assertion was tested by undertaking a simulation wherein the rate of technical progress was increased from the original 0.12, in chapter 5,[13] to 0.14.

Much as expected, the results show that domestic investment has been increased in most periods, with the overall result that the size of the terminal domestic capital stock is much larger than that shown in Table 6.3a where the rate of technical progress assumed is lower. We can also observe that during the planning period, a higher value of overseas assets is accumulated before funds are repatriated from overseas. The portfolio of assets peaks at a much higher level in 1988, here, compared with that of the base case for the same year in chapter 6. This is because the higher income generated as a result of the higher productivity, caused, in turn, by a greater level of technical progress, allows more investment overseas. The values of domestic non-oil income compared with that of Table 6.3c are much higher.

Finally, the time path of oil depletion is significantly different from that shown for a lower rate of technical progress in Table 6.3b. Production is brought forward considerably into the pre-oil-peak period, with a higher peak than that of chapter 6. During the 1990s the oil extraction rate is much lower than that of the basic case. The economic implication of this result is that when a higher rate of technical progress is envisaged, the oil resource can be depleted earlier without having to wait for higher oil prices in the later years. The additional income from domestic higher productivity will compensate for the higher oil prices in the later years.

7.9 Changes in the Rate of Return Overseas

It has already been stated that one of the crucial relationships in the model is that between the rate of return generated by the domestic production function and the rate of return generated by the overseas asset function. In the previous section we examined the effect of altering the former by varying the rate of technical progress. Here, the sensitivity of the model solutions to a change in the latter will be investigated.

The parameter signifying the yield on overseas investment is θ, as shown in equation 2.8 of the macromodel[14] reproduced below:

$$F_t = \theta E_{t-1}$$

In the base case this parameter was set at 5% per annum to represent the average real rate of return in the long term outside the United Kingdom. Here it is reduced first to 4% and then increased to 6%. However, to avoid repetition, we only discuss the results related to the case when θ is reduced to 4%. The results of the simulation when θ is increased to 6% are the reverse of the reduction to 4%. The same economic reasoning for the reduction can be applied to analyse and explain the effect of the increase in θ.

Our solutions show that when the real rate of return overseas is lower (compared to the case where the real rate of return abroad is five per cent as applied in chapter 6), the portfolio of assets overseas over the planning horizon becomes much slimmer.[15] This in turn leads to a lower global non-oil wealth accumulated over the life of oil, as shown by the value of the objective function.

The interesting features of these solutions are the optimal time paths of the control variables, oil depletion and non-oil domestic investment. Oil depletion is brought forward in time and the rate of extraction in the pre-oil-peak period (i.e., prior to 1988) is substantially higher than the basic case shown in Table 6.3b. Parallel to this, domestic investment is also much higher than that of the basic case (Table 6.3b) during the pre-oil-peak period. This poses the question of why a higher rate of depletion in the earlier years is optimal, in the face of a lower rate of return overseas? The answer in this case lies in the time path of the rate of return at home relative to that abroad.

The yield on domestic investment is that generated by the set of parameters of the production function postulated in chapter 5.[16] The underlying assumption there is that the production function exhibits decreasing returns to scale with respect to capital stock as a factor input. Further details were given in chapter 5.[17] It was evident in the solutions that as more investment takes place, the increments on the income generated gradually fall. Thus the real rate of return on home investment becomes higher in the 1980s decade than the 1990s. This is the main factor responsible for bringing forward the oil depletion into the 1980s.

Since F' is postulated as constant over the planning period (because an average long-term rate is considered), the key variable influencing the rate of petroleum extraction becomes D'. As a higher rate of return on investment can be fetched in the 1980s than in the 1990s decade, more extraction in the former years becomes viable, which leads to higher investment for these years. Thus we can see the reasoning behind the enhanced size of the domestic non-oil investment in the first decade of the planning period. When F' is projected at an average of 6% over the life of oil, it has just the opposite effect to the above. However, we shall not enter further into details of the numerical solutions.

7.10 Changes in the Stock of Oil Reserves in the North Sea. Variations in the stock of oil reserves in the North Sea can occur, and there is of course a great deal of uncertainty as to the exact magnitude of the recoverable oil lying in the UK Continental Shelf. The effect of changes in the stock of extractable oil on our planning model can be studied in two ways. One would be to consider larger oil reserves over the same planning horizon—that is, to the end of this century. The other would be to consider larger oil stocks over an extended period—the planning horizon would be extended as a result of the new discoveries. Here we have opted for the first alternative, mainly because the underlying framework for all the sensitivity analyses of this chapter has been to change one variable at a time and study its effect under *ceteris paribus* conditions.

For the purpose of this simulation, the stock of proven oil reserves is raised from its original value, of about twelve million barrels, to sixteen-and-a-half million barrels, to be depleted over an equal number of years. The solutions in this case indicate a much higher rate of oil extraction during the years 1988 – 92, which falls in the second half of the planning period. As one would expect, the enhanced oil reserves result in a much higher value of the objective function. However, an interesting feature of this simulation result is that a substantially larger portfolio of assets overseas is accumulated. The reason for this result is associated with the finite nature of the oil revenue. As a large oil stock is depleted, a large gap would have to be filled both in terms of import substitution and export earnings. Thus a higher level of foreign earnings would be required to compensate for the loss of oil revenue.

8 Conclusions

This volume has been based on the findings of a research project[1] which set out to examine the macroeconomics of North Sea oil using Optimal Control Theory. More specifically, the study has focused on depletion strategies in conjunction with expenditure policies of the resource's revenue. Such policies have been determined in relation to a basic objective, posed in the form of the following question: What is the maximum global non-oil wealth that can be accumulated over the life of the oil resource? The objective function is in fact interchangeable with maximizing aggregate consumption or aggregate income.[2] A long-run macroeconomic model of the UK is developed, and then combined with the technique of optimal control.

It should be stressed that while the UK economy is used here as an example of an economy with the North Sea oil as a non-replenishable resource, the ultimate aim of this research is to put forward a methodology in macroeconomic planning that can be applied to any economy with an exhaustible resource. Naturally, the model would be adapted to suit the specific characteristics of the country in question. Nevertheless, the methodology would be valid as a useful planning tool for any economy concerned with determination of resource depletion and investment policies at macroeconomic level.

This concluding chapter will specify the broad outlines and implications of the analysis of the text. Details of the techniques and methodology employed are explained as they occur in the text, and will not be restated here. Moreover, we shall dispense with discussing the intricacies of the parameter estimation and the general quantifications of the macroeconomic system. Our main concern will be with summarizing the overall approach and the policy implications that emerge when the model is put to the test.

Our planning exercise commences with the formulation of a macroeconomic model of the UK which portrays the long-run characteristics of the economy. The model is developed in such a way as to reveal the role of the newly-emerged oil sector in Britain both through the income it generates and the balance of payments effects it causes. Subsequently the objective function is stated, and some inter-temporal constraints concerning the balance of payments, in the form of a minimum imports requirements, and the stock of proven oil reserves are specified. The system is then optimized over time subject to these constraints and the macroeconomic structure. In this planning model we have chosen two main control variables, namely, the rate of oil extraction and the level of non-oil investment during that planning period which coincides with the life of oil.

The conceptual formulation of the model and the derivation of optimum choice over time are presented in chapter 2. This leads onto a set of analytical results which can be summarized as follows:

When the balance of payments constraint is non-binding over a certain period of time, condition (a), equalizing rates of return at the margin between home and overseas investment is the optimal policy. At the same time, under these conditions the rule for the production of oil should be according to a modified version of the Hotelling's rule. The optimal depletion rate calls for the rate of change of oil prices over time to be less than the rate of interest. It is less by the value of a fraction which consists of a combination of some parameters and the yield on overseas investment. The parameters comprise average propensity to consume, income tax rate and expenditure tax rate. Effectively, the combination of these parameters acts as some form of saving propensity which takes into account tax leakages. Thus, depending on the magnitude of these parameters, the rate of return on overseas investment and the rate of change of the price of oil, the depletion of petroleum is determined.

When the constraint first becomes binding, condition (b), this implies that imports have to be kept at their minimum level. The optimal solution calls for investment overseas and a cutback on oil production. This is because, given the structure of the economy, if under these circumstances investment at home takes place, more imports will be required. The higher volume of imports would not be offset by the non-oil exports alone, due to the high exchange rate, and most of the oil earnings of foreign exchange would have to be used up to pay for imports. Moreover, the economy would be put on an expansion path such that when the oil runs out and the foreign exchange dwindles, there would emerge a gap between the receipts and the payments.

When the balance of payments constraint is binding over consecutive periods, condition (c), the solution is indeterminate, as far as the allocation of investment between home and overseas is concerned, and one has to resort to the next stage which gives a clear-cut direction for policy. However, with respect to the depletion rule we obtain a conclusive solution, and that is: cut back on oil extraction. In other words, when the balance of payments is binding over a prolonged period, the optimal policy is to reduce petroleum production, as this will in turn reduce the exchange rate and help improve non-oil exports.

Finally, when the constraint is being removed, condition (d), the optimal policy calls for more investment at home, even if at a lower rate of return than can be fetched overseas, accompanied by a reduction on petroleum extraction. The rationale for domestic investment is to boost non-oil exports and to gain other sources of foreign exchange. At this stage the economy has to plan to reduce its dependence on petroleum, and although import requirements will rise as a result of high domestic investment, by this time the income from overseas assets would have sufficiently grown to offset the dwindling oil revenue. Investment in the non-oil exports at home during condition (c) will ensure a smooth transition into a state with a replenishable source of income.

While these results reveal the analytical aspects of the workings of the system and the direction of the policy decisions, the exact magnitude of all the variables can only be determined once the model is fully quantified. This is the subject of chapter 5, where the parameter values are estimated and the time-path of the exogenous variables is projected.

Prior to the quantification of the model, a survey of existing macroeconomic models of the UK is given in chapter 3. The purpose of this survey was twofold. Firstly to show that none of the existing macromodels of UK dealt with the question of North Sea oil depletion, and invariably treated it as exogenous—and hence to develop our own model, here. Secondly to highlight the areas of relevance of some of the equations of these models for the purpose of quantification of our model parameters.

In chapter 4 we give an account of all aspects of the North Sea oil sector in Britain. The structure of the oil taxation in Britain, as at the end of March 1982, is explained in detail. We look at the information available on planned production in the UK, tentative as it may be, as projected by the Department of Energy and the oil companies. This information is used to compute the likely magnitude of the oil revenue up to the end of this century. Simulations are carried out against variations in the level of projected oil prices, projected exchange rate, projected rate of oil extraction and projected inflation rate.

Having examined in depth all the relevant information available in the UK economy both on the modelling side and on the basic data for the quantification of our parameters and projections, the model is put to the test in chapter 6. Numerical solutions are obtained for all the variables in question. These first sets of results are defined as our basic solutions, against which sensitivity analyses are conducted in chapter 7.

The numerical solutions shown in chapter 6 essentially verify the analytical solutions put forward in chapter 2. They show clearly how the optimal decision would be to cut back on oil production when the balance of payments constraint is binding. The optimal rate of oil extraction turns out to be lower in the 1980s decade than those projected by the UK Department of Energy or various oil companies. Moreover, peak oil production in our optimal solutions occurs at a later date (in the late 1980s) than that projected by the Department of Energy or the oil market. It should be pointed out, of course, that the British government maintains that it has no specific policy with regard to the rate of North Sea oil extraction. Such figures as are released by the Department of Energy are mere projections and are subject to variations depending on economic circumstances.

The numerical solutions show that for the particular macroeconomic structure (i.e., with those sets of parameters and projected time paths of the exogenous variables) the point in time at which the constraint becomes binding marks the division of the planning period into two distinct sub-planning phases. These phases signify a clear strategy of accumulation of assets abroad in the 'former' years of the oil life and expansion of the domestic sector in the 'latter' years of the resource.

An interesting feature of the solutions is that the 'former' and the 'latter' sub-planning phases coincide with the 'pre-peak' and 'post-peak' of the optimal oil extraction profile. This building up of a portfolio of assets overseas becomes synonymous with the pre-oil peaks phase, and a rise in domestic investment combined with repatriation of funds from abroad for the growth of the home sector is concomitant with the post-oil peak phase of the resource.

Chapter 7 contains a sensitivity analysis around the parameter values and projected path of the exogenous variables which were used in the derivation of the solutions in chapter 6. A number of simulations are carried out and we shall briefly restate some of the results here.

We first removed the minimum production level imposed on the first four years of North Sea oil production (1976 – 79 inclusive). These constraints were imposed in chapter 6 to reflect the case that the operating companies had to recover their investment as fast as possible and were lifting the maximum amount of crude oil that was technologically possible. The result of this simulation shows that the removal of this constraint radically changes the oil depletion path in the first four years. The optimal rate of oil depletion turns out to be much lower than the actual rate. This indicates that the rate of oil extraction in the model is quite sensitive to oil prices, since these fell in real terms during the later 1970s.

When a general rise in the level of government expenditure over the entire life of North Sea oil is envisaged, the simulation result shows a decline in the level of non-oil wealth accumulated during this period. The time profile of optimal oil depletion remains on the whole fairly robust, although production is marginally brought forward into the first half of the 1980s to finance the increased government expenditure. In the later years the domestic non-oil income will finance the additional government expenditure, which implies that less funds will be allocated to investment.

In a situation where a higher level of aggregate consumption is visualized, domestic investment becomes reduced with a consequent decline in the accumulated value of non-oil wealth. The overall pattern of the optimal oil extraction rate again remains fairly robust, although the oil production rate is stepped up a little during the pre-oil peak years. The rationale for these results is similar to the case where a higher government expenditure is introduced.

Variations in the rates of taxation, both in the income related taxes and expenditure taxes, have been the subject of some of the other simulations. Since government consumption expenditure is treated as exogenous in the model, changes in taxation would affect the funds available for the public sector capital expenditure. The results of a cut in taxation rate are similar to those of a higher government expenditure, or higher consumption, although the exact values of all the variables would depend on the magnitude of the change to which the system has been subjected.

On the balance of payments side, sensitivity analysis is carried out the effects of oil revenue on the level of effective change rate. When a weaker response of the exchange rate to oil revenue is postulated, the optimal rate of oil extraction becomes reduced in the 1980s and enhanced in the 1990s. The alteration in the path of oil depletion is due to the effect of exchange rate on non-oil exports. As the influence of the oil revenue on the exchange rate deteriorates, non-oil exports do not suffer as much, and remain more competitive in the international market. Hence oil extraction does not have to rise by as much in the earlier years to compensate for the loss in non-oil exports. Higher oil production can be left to the later years when the price of oil is higher. Furthermore, when the oil revenue is postulated to have a weaker impact on the exchange rate, it implies that the value of overseas assets in sterling terms is not eroded by as much. Hence, the portfolio of assets abroad remain larger when assessed in sterling currency.

As far as the domestic non-oil production is concerned, when a higher rate of technical progress is conceived, oil production is brought forward in time, accompanied by higher domestic investment. The reasoning behind this is that the additional income from the increased productivity at home will compensate for the higher oil prices likely to occur in the later years of the life of the resource.

On the overseas income side, when the rate of return abroad is lowered, the rate of oil extraction is again stepped up during the pre-oil peak period. The cause of this alteration in the time path of oil depletion here is somewhat different from the above cases. It is associated with the decreasing returns-to-scale nature of the domestic non-oil output function, with respect to capital input.[3] This implies that the marginal product of capital tends to be higher in the pre-oil peak period, generating a higher marginal rate of return in that phase. Hence, high oil production is called for in that period, which will in turn lead to a higher non-oil wealth accumulated over the life of oil.

Two more sets of simulations specifically related to the oil sector are conducted. One is related to the forecasts of oil prices, and the other pertains to likely changes in the stock of oil reserves. Different scenarios are drawn up for the likely changes in the price of oil and it is shown how the oil depletion path is affected. When a high rate of price change is conjectured for the future, oil production is deferred to the future, the 1990s. The opposite is true when a low rate of price change is stipulated.

With regard to variations in the stock of oil reserves in the North Sea, a simulation is conducted for an increased stock of oil as a result of new discoveries. The solutions show a much higher rate of extraction occurring in the years 1988 – 92, which falls in the second half of our planning period. The additional oil reserves, as one would expect, result also in a much higher value of the objective function, namely the accumulated non-oil wealth by the time the oil resource is exhausted.

Finally, it should be stressed that since the prime purpose of this study is to demonstrate a methodological framework for planning, the emphasis should be placed more on the relative rather than the absolute values of the numerical solutions for all the macroeconomic variables. Indeed, it is the direction and timing of these variables that would be of interest to the policy-maker. Such a planning model would be capable of providing this information. Needless to say, the research undertaken for this study is not by any means finite. One would ideally wish to introduce more refinement and sophistication into the model. Nevertheless, although what is presented here are the bare bones of the model structure, its foundation has been laid such that it can cater for further disaggregation and incorporation of different viewpoints.

1 References

1 Introduction

1 In Spring 1978 the Times newspaper ran a debate on how the advantages which would flow from the North Sea should best be exploited in the interests of the British people as a whole. For an example of some of the more specific suggestions put forward by the public see the letter by Lord Seebohm, *The Times*, 12 April 1978.

Another debate was run in 1980 (during August and December) in the *Guardian* under the title 'Does our oil really make us richer?' and the main contributors were T. Barker, J. Kay, P. Forsyth and F. Edwards.

2 Government White Paper, 'The Challenge of North Sea Oil', March 1978.

3 P.O. Forsyth and J.A. Kay, 'The Economic Implications of North Sea Oil Reserves', *Fiscal Studies*, Vol. 1 (1980), No. 3, pp. 1–28.

4 The notion of a binding balance of payments constraint refers to an economic condition (defined here) where the level of imports cannot be reduced any further.

5 This holds true, of course, at the time this research was completed. It may well be that some of these models will be expanded to endogenize the oil sector. However, so far the oil sector has been treated as exogenous in these models.

6 See Table 5.14.

7 'Basic' model results refer to the numerical solutions of chapter 6, using the parameters estimated in chapter 5.

2 The Model and Its Analytical Solutions

1 Some would argue that a better term to describe the behaviour of the UK is price-follower. Once the OPEC countries announce their price for a comparable crude to that of the UK (mainly Nigeria and Libya), Britain adjusts her prices accordingly shortly thereafter. Either term would nevertheless confirm the assumption of exogeneity about the price of oil in the British economy.

2 This is the reason for expressing P_t at market prices in equation 2.1 above, for it includes l_t^1, the tax on petroleum output.

3 Supplementary petroleum duty was a category of oil taxation until the end of 1982.

4 This is a rather crucial variable in the model, for it is the spending of oil proceeds that is the concern of this study.
5 The first partial derivative of non-oil output, D_t, in equation 2.2 in the text with respect to the lagged value of capital stock, K_{t-l}, determines the incremental output-capital ratio outside the petroleum sector.
6 Our definition of income-related-taxes includes direct income tax, corporation tax and national insurance.
7 For a general application of the Kuhn-Tucker Theorem see K.J. Arrow and M. Kurz, *Public Investment, the Rate of Return and Optimal Fiscal Policy,* (Baltimore: Johns Hopkins University Press, 1971), pp.26 – 55, and M.D. Intriligator, *Mathematical Optimization and Economic Theory*, (Englewood Cliffs, N.J.: Prentice Hall, 1971), pp. 292 – 3.
8 In order to arrive at solutions to our optimal control problem, we need to derive the Kuhn-Tucker conditions to our Lagrangian (2.27).
9 See Intriligator, op.cit. pp.292 – 3.
10 For a more elaborate analysis of the derivation and implications of such conditions see H. Motamen, *Expenditure of Oil Revenue* (St. Martin's Press, New York, and Frances Pinter, Publisher, London, 1979), pp. 30 – 9.
11 H.Hotelling, 'The Economics of Exhaustible Resources', *Journal of Political Economy*, Vol. 39 (1931), pp.137 – 75.

3 Survey of the Principal UK Macroeconomic Models

1 D. A. Livesey, solving the model, in *Economic Structure and Policy with Applications to the British Economy*, pp 54 – 86. Edited by Terence S. Barker. Cambridge Studies in Econometrics No. 2 (London: Chapman & Hall, 1976).
2 W. W. Leontief, *Input-Output Economics*, (New York: Oxford University Press, 1966).
3 R. J. Ball and T. Burns, the Inflationary Mechanism in the UK Economy. *American Economic Review*, Vol. 66, No. 4 (September 1976), pp.467 – 84.
4 P. A. Ormerod, the National Institute Model of the UK Economy: Some Current Problems, in *Economic Modelling*. Edited by P. A. Ormerod. (London: Heinemann Educational Books, 1979).

4 Taxation of UK Oil and Estimation of Oil Revenue

1 Continental Shelf Act (London: HMSO, 1964).
2 See *Economic Progress Report*, No. 131 (London: HMSO, March 1981) pp. 12 – 14.
3 See *Economic Progress Report*, No. 143 (London: HMSO, March 1982) pp. 4 – 5.
4 Oil Taxation Act (London: HMSO, 1975).
5 See *Economic Progress Report*, No. 143 (London: HMSO, March 1982) pp. 4 – 5.
6 See *Economic Progress Report*, No. 101 (London: HMSO, August 1978) p. 1.
7 Finance Act (London: HMSO, 1979).
8 See *Economic Progress Report*, No. 116 (London: HMSO, December 1979) p. 1.

9 See *Economic Progress Report*, No. 120 (London: HMSO, April 1980) p. 8.
10 See *Economic Progress Report*, No. 131 (London: HMSO, March 1981) pp. 12 – 14.
11 It is now believed that the Cormorant field, instead of consisting of two distinct reservoirs—South Cormorant and North Cormorant— is in fact contiguous and should, therefore, be treated as a single field for taxation purposes.
12 The figure of 11,949 million barrels refers to an estimate— current during mid-1981—of ultimate recoverable reserves from the aforementioned fields. Estimates of recoverable reserves and depletion rates for individual fields are, however, regularly subject to considerable revision and, for the purposes of these calculations, more recent forecasts have been used. These predict both a lower level of recoverable reserves and also depletion from some fields continuing into the twenty-first century. Hence, the level of oil production expected during the period 1976 – 2000 is forecast here at only 11,431 million barrels.
13 Petroleum and Submarine Pipe-Lines Act (London: HMSO, 1975).
14 Gearing denotes the proportion of funds which are externally financed.
15 For an analysis of the effect of different tax regimes—up to and including that incorporating the changes announced in the course of the 1980 Budget—see the article by Homa Motamen and Roger Strange, Oil Revenue Outlook for Britain in the Medium Term, *Energy Policy*, Vol. 9, No. 1 (March 1981), pp. 14 – 19.
16 11,949 million barrels corresponds to the initial oil stock assumed in the calculations of chapters 5, 6 and 7, which was a current mid-1981 estimate of recoverable reserves.

5 Quantification of the Model Parameters and Projections of the Exogenous Variables

1 A good elucidation of the main issues is provided by David F. Heathfield, 'Production Functions', in *Topics in Applied Macroeconomics*, pp. 33 – 68. Edited by David F. Heathfield (London: Macmillan, 1976). See also A. A. Walters, Production and Cost Functions an Econometric Survey, *Econometrica*, Vol. 31 (1963), pp. 1 – 66.
2 For a further exposition of the properties of transcendental logarithmic functions, see the two articles by Lavrito R. Christensen, Dale W. Jorgenson and Lawrence J. Lau:
(i) Conjugate Duality and the Transcendental Logarithmic Functions, *Econometrica*, Vol. 29, (1971), pp. 255 – 6.
(ii) Transcendental Logarithmic Utility Functions, *American Economic Review*, Vol. 65, (1975), pp. 367–83.
3 Rita Maurice, (ed.), *National Accounts Statistics: Sources and Methods* (London: HMSO, 1968) p. 14.
4 See, for example, G. C. Harcourt, *Some Cambridge Controversies in the Theory of Capital* (London: Cambridge University Press, 1972).
5 Tom Griffen, The Stock of Fixed Assets in the United Kingdom: How to Make Best Use of the Statistics, *Economic Trends*, No. 276 (London: HMSO, October 1976) p. 136.

6 *Ibid*. p. 133.
7 The 1980 edition of the *National Income and Expenditure* has been used as the main source for our statistics in this chapter.
8 See 'Companies Rate of Return on Capital Employed: 1960-1979', *British Business*, Vol. 3, No. 5 (3 October 1980), pp. 222 – 3.
9 See Table 2.1.
10 See reference 8.
11 See Table 2.2.
12 Rita Maurice, (ed.), op.cit., p. 448.
13 See p. 14 above.
14 Samuel Brittan, 'Real' Reasons for High Interest Rates, *Financial Times* (17 September 1981), p. 23.
15 Simon Kuznets, *National Product since 1869* (New York: National Bureau of Economic Research, 1946).
16 Milton Friedman, *A Theory of Consumption Function* (Princeton, N.J.: Princeton University Press, 1957).
17 Simon Kuznets *op. cit.*
18 Milton Friedman *op. cit.*
19 See Terence S. Barker, *Economic Structure and Policy: with Applications to the British Economy*, (London: Chapman & Hall, 1976).
20 James E. H. Davidson; David F. Hendry; Frank Srba and Stephen Yeo, Econometric Modelling of the Aggregate Time-Series Relationship between Consumers' Expenditure and Income in the United Kingdom, *Economic Journal*, Vol. 88 (1975), pp. 661 – 92.
21 See M. J. Fetherston and K. J. Coutts, *Technical Manual on the CEPA Model of the UK*, 4th edition (Cambridge: Cambridge Economic Policy Group, May 1979), pp. 14 – 16.
22 See Table 2.2.
23 Yoon S. Park, *Oil Money and the World Economy* (London: Wilton House, 1976) pp. 25 – 43.
24 *National Income and Expenditure: 1980 Edition* (London: HMSO, 1980) p. 17.
25 See Section 5.2.
26 See Section 2.3. The reduced form for the E_t equation is 2.20 shown in chapter 2.
27 *National Income and Expenditure: 1980 Edition* op. cit, p. 84.
28 *UK Balance of Payments: 1980 Edition* (London: HMSO, 1979) p. 11.
29 *Development of the Oil and Gas Resources of the United Kingdom* (London: HMSO for the Department of Energy, 1980) p. 5.
30 Continental Shelf Act (London: HMSO 1964).
31 See reference 29.
32 Wood, Mackenzie, *North Sea Service: Field Analysis Section*, Bulletins 1979 to Aug. 1980. These bulletins contain estimates for different oilfields, and are continually updated.
33 *Development of the Oil and Gas Resources of the United Kingdom*, op. cit. p. 3.
34 *Ibid*, p. iv.

6 Numerical Solutions for the Model

1 The estimation of each of the parameters is explained in detail in chapter 5.
2 See section 5.11.
3 For further details of oil price behaviour, see section 5.10.
4 This information was given by Angus Beckett, in a seminar at the University of Surrey in May 1979.
5 See Table 2.2.
6 See Table 4.4.
7 These are the values of the macroeconomic variables as they appear in the *National Income and Expenditure* statistical book. Further details are shown at the foot of Tables 6.4 and 6.5.
8 See section 6.3.
9 See section 2.5.
10 *Ibid.*
11 These optimal solutions emerge as a result of the simultaneous solution of the model for the entire length of the planning horizon. Thus optimal level of investment is determined with the foresight of the constraint value on the subsequent year.
12 The 'Brown Book' is the general name for the publication which appears annually under the title *Development of Oil and Gas Resources of the United Kingdom*. It is published by HMSO.
13 See *Development of Oil and Gas Resources of the United Kingdom, 1981* (London: HMSO 1981), p. 54.
14 See pp. 38–45.
15 See A. G. Kemp and D. Crichton (1979) Effects of Changes in UK North Sea Oil Taxation *Energy Economics*, Vol. 1, No. 4, October.
16 Colin Robinson and Chris Rowland have published different papers on this subject. The most relevant paper here would be: Rowland, C., Taxing North Sea Oil Profits in the U.K.: Special Needs and the Effects of Petroleum Revenue Tax, *Energy Economics*, Vol. 2 (April 1980), pp. 115 – 25.
17 There could of course be other projections made on the future rate of North Sea oil depletion which were not known to us. We did try to cover the literature on this subject as thoroughly as we could, and British Petroleum and Wood Mackenzie were the only two sources that served as a basis of comparison.
18 In fact all the field-by-field analyses of Kemp and Crichton, *op. cit.* used Wood Mackenzie's figures as a basis for their calculations.
19 UKCS stands for United Kingdom Continental Shelf.
20 The oil peak of 2.6 million barrels a day in 1988 may seem, *prima facie*, a sharp rise compared to the previous year. Petroleum engineers claim such a rise could be absorbed if it were planned well in advance, and it should be borne in mind that the emphasis here is on long-range planning. Moreover, the optimal oil depletion path generated here is unconstrained and if a sharp rise technologically is feasible, additional constraints can always be imposed to further smooth out the path. However, it is not necessary in this case.
21 The fall in the oil extraction rate in 1989 may appear a little too drastic. The magnitude of this cutback was discussed with various reservoir and petroleum engineers at London University's Imperial College.

They confirmed that a reduction in oil production for a brief period, of, say, a couple of years, would be geologically feasible.

22 The British Petroleum figure for North Sea oil reserves based on existing fields and fields under development comes to 10.939 million barrels. This figure is estimated here by multiplying their projected values of daily extraction by 365 and summing the annual values over the period 1976 – 2000.

23 Peak oil production occurs in 1988 according to the optimal policies derived here.

24 See Table 2.2.

25 Technology may accelerate the development of other sources of energy, and the need for petroleum imports may disappear. However, given the short horizon considered for the life of North Sea oil (two-and-a-half decades), and given the various uses of petroleum as an energy input, it may well turn out that oil would have to be imported.

26 See Table 2.2.

7 Simulations and Sensitivity Analyses

1 For the real price of oil during 1976 – 80 see Table 5.13.

2 See section 6.3.

3 As shown in Table 6.3b.

4 For details of the items that are included in the categories of 'income-related' and 'expenditure' taxes see section 5.6.

5 Our definition of government expenditure, G_t is limited to the consumption expenditure of the public sector. For more details see chapter 5, p. 94.

6 See table 2.2.

7 See section 5.7.

8 This applies to the period after the initial four years when the minimum extraction constraint which we imposed is binding. After the first four years the rate of oil extraction is lower with the reduced value of G_t.

9 See Table 2.2.

10 See p.88.

11 See p.86.

12 The production function in the macroeconomic model in chapter 2 is shown as equation 2.2. See Table 2.2.

13 See p.80.

14 See Table 2.2.

15 See Table 6.3a.

16 See chapter 5, pp. 85–8.

17 *Ibid.*

8 Conclusions

1 See Preface.

2 See Chapter 2, pp. 15–16.

3 *Ibid.*, p. 8.

Bibliography

Arrow, K.J. and Kurz, M. (1971) *Public Investment, the Rate of Return and Optimal Fiscal Policy*, Baltimore: John Hopkins University Press.
Attanasi, E.D. (1979) the nature of firm expectations in petroleum exploration, *Land Economics*, Vol. 55.
Ball, R.J; Boatwright, B.D; Burns, T; Lobban, P.W.H. and Miller, G.W. (1975) The London Business School Quarterly econometric model of the UK economy, in *Modelling the Economy*, Edited by G.A.Renton, London: Heinemann Educational Books.
Ball, R.J. and Burns, T. (1976) the inflationary mechanism in the UK economy, *American Economic Review*, Vol. 66, No. 4, September.
Barker, T.S. (1976) *Economic Structure and Policy: with Applications to the British Economy*, Cambridge Studies in Applied Econometrics, No. 2, London: Chapman Hall.
———— (1980) Oil-funded Investment is the only way to invigorate Industry, *Guardian*, 18 August.
Beudell, M. (1976) European offshore: tax and price effect on oil output, *Petroleum Economist*, Vol. 43, December.
British Business (1980) 'Companies rate of return for capital employed: 1960 – 1979', Vol. 3, No. 5, 3 October.
Brittan, S. (1981) 'Real' reasons for high interest rates, *Financial Times*, 17 September.
Cairncross, F. (1980) How do we get Sterling down if we can't persuade the Market we're on our uppers?, *Guardian*, 3 May.
———— (1980) Taxman sets a poverty trap for North Sea oil explorers, *Guardian*, 20 May.
Christ, C.F. (1975) Judging the performance of econometric models of the US economy, *International economic Review*, Vol. 16.
Christensen, L.R; Jorgensen, D.W. and Lau L.J. (1971) Conjugate duality and the transcendental logarithmic function, *Econometrica*, Vol. 29.
———— Transcendental logarithmic utility functions, *American Economic Review*, Vol. 65.
Continental Shelf Act (1964) London: HMSO.
Council for Science and Society (1979) *Deciding about energy policy; principles and procedures for making energy policy in the UK*, London: The Council for Science and Society.
Coutts, K.J. (1977) Post mortem on five years of CEPG forecasting Bib.2, *Economic Policy Review*, No. 3.

Coyne, T.J. (1980) Oil industry profitability: an interindustry comparison of returns to equity, *Business economics*, May.
Cripps, F. and Tarling, R. (1975) An economic assessment of the North Sea, *Economic Policy Review*, No. 1, February.
Davidson, J.E.H; Hendry, D.F; Srba, F. and Yeo, S. (1975) Economic modelling of the aggregate time-series relationship between consumers' expenditure and income in the United Kingdom, *Economic Journal*, Vol. 88.
Davidson, P. (1979) Oil Conservation: theory vs policy, *Journal of Post-Keynesian Economics*.
Department of Energy (1974) *Production and Reserves of Oil and Gas in the United Kingdom*, A report to Parliament by the Secretary of State for Energy, May, London: HMSO.
———— (1974) *United Kingdom Offshore Oil and Gas Policy*, Cmnd 5695, London: HMSO.
———— (1976) *Nationalised Industries and the Exploitation of North Sea Oil and Gas*, Cmnd 6408, London: HMSO.
———— Study group (1976) *North Sea Costs Escalation Study*, Part 1: Department of Energy Study Group, Part 11: Peat Marwick Mitchell and Co. and Atkins Planning, Energy Paper No. 7, London: HMSO.
———— (1976) *Development of the Oil and Gas Resources of the United Kingdom*, A Report to Parliament by the Secretary of State for Energy, April, London: HMSO.
———— (1977) *Development of the Oil and Gas Resources of the United Kingdom*, A Report to Parliament by the Secretary of State for Energy, April, London: HMSO.
———— (1978) *Development of the Oil and Gas Resources of the United Kingdom*, a report to Parliament by the Secretary of State for energy, April, London: HMSO.
———— (1978) *The Challenge of North Sea Oil*, Cmnd 7143, March, London: HMSO.
———— (1980) *Development of the Oil and Gas Resources of the United Kingdom*, a Report to Parliament by the Secretary of State for Energy, May London: HMSO.
Economic Progress Reports (1978 to 1982) London: HMSO.
Ericson, J. (1980) Who will control North Sea oil?, *Marxism Today*, Vol. 24.
Fetherston, M.J. (1976) *Technical Manual on the CEPG Model*, Cambridge: Department of Applied Economics.
———— and Coutts, K.J. (1979) *Technical Manual on the CEPG Model of the UK*, May, Cambridge: Cambridge Economic Policy Group.
Field, B.C. and Grebenstein, C. (1980) Capital-energy substitution in US manufacturing, *Review of Economics and Statistics*, Vol. 62.
Finion, D. (1979) Scope and limitations of formalized optimization of a national energy system: the EDOM model, in *Energy Models for the European Community*, Edited by A. Strub, Guildford: IPC Science and Technology Press Ltd for the Commission of the European Communities.
Forsyth, P.J. and Kay, J.A. (1980) The economic implications of North Sea oil revenues, *Fiscal Studies*, Vol. 1, No. 3.
Friedman, M. (1957) *A Theory of Consumption Function*, New Jersey: Princeton University Press.
Gregory, R.G. (1976) Some implications of the growth of the mineral sector, *The Australian Journal of Agricultural Economics*, Vol. 20.
Griffen, T. (1976) The stock of fixed assets in the United Kingdom: how to make best use of the statistics, *Economic Trends*, No. 276, October, London: HMSO.

Hamilton, A. (1978) North Sea Impact: *Off-Shore Oil and the British Economy*, London: International Institute for Economic Research.
——— (1980) The North Sea billions, *Observer*, 24 February.
Harbaugh, J.W; Doveton, J.H. and Davis, J.C. (1977) *Probability Methods in Oil Exploration*, New York: John Wiley.
Harcourt, G.C. (1972) *Some Cambridge Controversies in the Theory of Capital*, London: Cambridge University Press.
Harris, D.J. and Davies, B.C.L. (1980) Resource allocation in the nationalized energy sector of the United Kingdom, *Managerial and Decision Economics*, Vol. 1, No. 1.
Hayllar, R.F. and Pleasance, R.T. (1977) *UK Taxation of Offshore Oil and Gas*, London: Butterworths.
Heathfield, D.F. and Pearce, I.F. (1975) A view of the Southampton econometric model of the UK and its trading partners, in *Modelling the Economy*, Edited by G.A. Renton, London: Heinemann Educational Books.
Heathfield, D.F. (1976) Production functions, in *Topics in Applied Macroeconomics*, London: Macmillan.
Hotelling, H. (1931) The economics of exhaustible resources, *Journal of Political Economy*, Vol. 39.
Houthhakker, H.S. (1980) The use and management of North Sea oil, in *Britain's Economic Performance*, Edited by Richard E. Caves and Lawrence B. Krause; Comments by W. Maynard and Michael V. Posner, Washington D.C.: Brookings Institution.
Input-Output Tables for the United Kingdom, Business Monitor, PA 1004, up to 1974, London: HMSO.
Intriligator, M.D. (1971) *Mathematical Optimization and Economic Theory*, Englewood Cliffs, N.J.: Prentice Hall.
Kaufman, G.M. (1975) Models and methods for estimating undiscovered oil and gas—what they do and don't do, Proceedings of the IIASA Conference on Assessing Energy Resources, October.
Kemp, A.G. and Cohen, D. (1980) The new system of petroleum revenue tax, *Fiscal Studies*, Vol. I.
——— (1979) Effects of changes in UK North Sea oil taxation, *Energy Economics*, Vol. I.
Kemp, A.G. and Van Long, N. (1980) On two folk theorems concerning the extraction of exhaustible resources, *Econometrica*, Vol. 48.
Kirby, M.A. (1979) How much oil is there?, *The Banker*, Vol. 129, No. 645.
Kurz, M. (1971) *Public Investment, the Rate of Return and Optimal Fiscal Policy*, Baltimore: John Hopkins University Press.
Kuznets, S. (1946) *National Product since 1869*, New York: National Bureau of Economic Research.
Latter, A.R. (1979) *Bank of England Model of the UK Economy*, Bank of England Discussion Paper No. 5, London: Bank of England.
Laury, J.S.E; Lewis, G.R. and Ormerod, P.A. (1978) Properties of macroeconomic models of the UK Economy: a comparative study, *National Institute Economic Review*, No. 82.
Leontief, W.W. (1966) *Input-Output Economics*, New York: Oxford University Press.
Lewis, R. (1979) The exhaustion and depletion of natural resources, *Econometrica*, Vol. 47.
Livesey, D.A. (1971) Optimising short-term economic policy, *Economic Journal*, Vol. 81.
——— (1973) Some further results for a model of the UK economy, Paper presented at the IFACL IFORS Conference on Dynamic Modelling

and Control of National Economies, University of Warwick, 9 – 12 July.
———— (1976) Solving the model, in *Economic Structure and Policy with Applications to the British Economy*, Edited by T. S. Barker, Cambridge studies in Applied Econometrics, No. 2, London: Chapman & Hall.
London Business School (1979) *The London Business School Quarterly Econometric Model of the United Kingdom Economy: Relationships in the Basic Model as at September 1979*, London: London Business School.
Manser, W.A.P. (1979) Oil: a curious crisis, *The Banker*, Vol. 129, September.
Maurice, R. (ed.) (1968) *National Accounts Statistics: Sources and Methods*, London: HMSO.
Miller, M. (1980) Inflexible monetary targets could mean a 20 per cent loss in competitiveness, *Guardian*, 18 August.
Morgan, J.R. and Robinson, C. (1976) The comparative effects of the UK and Norwegian oil taxation systems on profitability and government revenue, *Accounting and Business Research*, Winter.
Motamen, H. (1979) Economic policy and exhaustible resources, OPEC Review, Vol. 111, No. 1, March.
———— (1979) Long-term objectives and planning of an oil-based economy, *Proceedings of International Symposia on Modelling, Planning and Decision in Energy Systems*, September. Zurich: IATA Publications.
———— (1980) *Expenditure of Oil Revenue: An Optimal Control Approach with Application to the Iranian Economy*, New York: St. Martin's Press, and London: Frances Pinter.
———— and Strange, R.N. (1980) UK government revenue from North Sea oil 1975 – 85, *The Banker*, Vol. 130, No. 653, July.
———— and Strange, R.N. (1981) Oil revenue outlook for Britain in the medium term, *Energy Policy*, Vol. 9, No. 1, March.
National Income and Expenditure: 1980 Edition (1979) London: HMSO.
National Institute of Economic and Social Research (1977) Some aspects of the medium-term management of the economy, *National Institute Economic Review*, No. 79, February.
———— (1977) Medium-term policy options, *National Institute Economic Review*, No. 82, November.
———— (1978) The medium term, *National Institute Economic Review, No. 86, November.*
———— *(1979) The Medium Term, National Institute Economic Review*, No. 90, November.
Norman, A.L. (1978) A first period certainty equivalence formulation of the MacRae open loop constrained variance strategy, Paper presented at the 7th NBER Conference on Economics and Control, Austin, Texas, 24 – 26 May.
———— ; Morris, R. and Palash, C.J. On the computation of deterministic optimal macroeconomic policy, The University of Texas at Austin, Department of Economics Discussion Paper, No. 74-1.
Odell, P.R. and Rosing, K.E. (1980) *The Future of Oil: A Simulation Study of the Inter-relationships of Resources, Reserves and Use, 1980—2080*, London: Kogan Page.
Ormerod, P.A. (1979) The National Institute Model of the UK Economy: Some Current Problems, in *Economic Modelling*, edited by P.A. Ormerod. London: Heinemann Educational Books.
Page, S.A.B. (1972) The value and distribution of the benefits of North Sea oil and gas, 1970 – 1985, *National Institute Economic Review*, No. 82, November.
Park, Y.S. (1976) *Oil Money and the World Economy*, London: Wilton

House.
Parkin, M. (1980) Oil push inflation?, *Banca Nazionale del Lavoro Quarterly Review*, No. 133.
Petersen, W. (1979) Fuel use in the UK: a study of substitution responses, in *Energy Models for the European Community*, Edited by A. Strub, Guildford: IPC Science and Technology Press Ltd., for the Commission of the European Communities.
Petroleum Economist (1979) North Sea revenue options, Vol. 46, July.
——— (1980) UK seventh round larger than expected, Vol. 47, June.
Petroleum and Submarine Pipe-Lines Act 1975 (1975) chapter 74, London: HMSO.
Pindyck, R.S. (1973) *Optimal Planning for Economic Stabilisation: The application of Control Theory to Stabilisation Policy*, Contributions to Economic Analysis No. 81, London: North-Holland.
——— (1973) Optimal policies for economic stabilisation, *Econometrica*, vol. 41.
Polak, E. (1971) *Computational Methods in Optimization: A Unified Approach*, New York: Academic Press.
——— (1972) A survey of methods of feasible directions for the solution of optimal control problems, *IEEE Transactions on Automatic Control*, AC-17.
——— ; Trahan R. and Mayne D.G. (1979) Combining phase I—phase II methods of feasible directions, *Mathematical Programming*, Vol. 17.
Posner, M.V. (1973) *Fuel Policy: A Study in Applied Economics*, London: Macmillan.
——— (ed.) (1978) *Demand Management*, National Institute of Economic and Social Research, Economic Policy Papers I, London: Heinemann Educational Books.
Quinlan, M. (1979) NOC's three years of progress, *Petroleum Economist*, Vol. 46, January.
——— (1979) UK North Sea: close to self-sufficiency, *Petroleum Economist*, Vol. 46, April.
——— (1979) Effects of UK oil revenues, *Petroleum Economist*, Vol. 46, July.
——— (1980) UK North Sea: self-sufficiency in Oil, *Petroleum Economist*, Vol. 47, June.
Ramsey, J.B. (1980) The economics of oil exploration: a probability-of-ruin approach, *Energy Economics*, Vol. 2.
Ray, G.F. (1977) The 'real' price of crude oil, *National Institute economic Review*, No. 82, November.
Reece, D.K. (1979) An analysis of alternative bidding systems for leasing offshore oil, *The Bell Journal of Economics*, Vol. 10.
Renton, G.A. (ed.) (1975) *Modelling the Economy*, London: Heinemann Educational Books.
Robinson, C. and Morgan, J. (1978) *North Sea Oil in the Future*, London: Macmillan Press for the trade Policy Research Centre.
Robinson, C. and Rowland, C. (1978) UK North Sea: marginal effect of PRT changes, *Petroleum Economist*, Vol. 45, December.
Robinson, J.N. (1980) North Sea oil and its uses: a good opportunity to link economic theory and practical matters, *Economics*, Vol. 16.
Rowland, C. (1980) Taxing North Sea oil profits in the UK: special needs and the effects of petroleum revenue tax, *Energy Economics*, Vol. 2, April.
Ryan, J.M. (1966) Limitations of statistical methods for predicting petroleum and natural gas reserves and availability, *Journal of Petroleum Technology*, Vol. 18.

Salomons, C.S. (1978) United Kingdom: the Oil Taxation Act 1975, *European Taxation*, Vol. 18.

Schwartz, A. and Regev, U. (1978) Estimation of production functions of specification bias due to aggregation, *Metroeconomica*, Vol. 30.

Seebohm, Lord (1978) Letter, *The Times*, 12 April.

Smith, V.K. and Krutilla, J.V. (1980) Toward a restructuring of the treatment of natural resources in economic models, Department of Economics Working Paper, No. 80 – 1, January, University of North Carolina.

Spencer, P; Mowl, C; Lomax, R. and Denham, M. (1978) *A Financial sector for the Treasury Model*, Part 1: *The Model of the Domestic Monetary System*, by P. Spencer and C. Mowl. Part 2: *The Model of External capital Flows*, by R. Lomax and M. Denham, Treasury Working Paper No. 8, London: HM Treasury.

Strub, A. (1979) *Energy Models for the European Community*, Guildford: IPC Science and Technology Press Ltd., for the Commission of the European communities.

Tempest, L.P. and Walton, R.T. (1979) North Sea oil and gas in the UK balance of payments since 1970, *Bank of England Quarterly Bulletin*, Vol. 19, September.

Tory Reform Group (1980) *Britain's Oil: Black Gold or Black Death*?, London: Tory reform Group.

United Kingdom Balance of Payments: 1980 Edition (1979) London: HMSO.

Wall, D. and Westcott, J.H. (1974) Macroeconomic modelling for control, *IEEE Transactions on Automatic Control* AC – 19.

Walters, A.A. (1963) Production and cost functions: an econometric survey, *Econometrica*, Vol. 31.

Wilson, T. (1979) The price of oil: a case of negative marginal revenue, *The Journal of Industrial Economics*, Vol. 27.

Wolf, C; Relles, D.A. and Navarro, J. (1980) *The Demand for Oil and Energy in Developing Countries*, Rand Report R-2488-DOE, Santa Monica: Rand corporation for the US Department of Energy.

Wood, Mackenzie (1979 – 1982) *North Sea Service: Field Analysis Section*, Wood Mackenzie and Co. Edinburgh.

Zoutendijk, G. (1970) *Methods of Feasible Directions*, Amsterdam: Elsevier.

Index